SIGNAL CHARLEY
FLAPS, WHEELS, HOOK DOWN

SIGNAL CHARLEY

FLAPS, WHEELS, HOOK DOWN

VIVID RECOLLECTIONS OF:

HARRY D. HAMILTON

Signal Charley

Copyright © 2005 by Harry D. Hamilton

Cover Image: Skyraider Coming Aboard (US Navy)
Interior Book Design: www.integrativeink.com

ISBN 1-4116-5508-7

To Jackie, my wonderful wife of fifty years, and counting.

US Naval Historical Center Photo of WWI Navy Recruiting Poster
(modified to b&w)

CONTENTS

Prologue ix

1 Midshipman Third Class Cruise 1

2 Midshipman Second Class Cruise 15

3 Midshipman First Class Cruise 21

4 Commission - Basic Flight Training 31

5 Advanced Flight Training 51

6 VAW-11 – Pre-Deployment 65

7 VAW-11 Cross-Country Flights 81

8 Survival School & Escape and Evasion 93

9 Carrier Operations & Deployment 101

10 VAW-11 – Post-Deployment 135

11 Naval Postgraduate School 153

12 FWC/JTWC Guam 157

13 FWC Suitland – Project FAMOS 171

14 Sweden: Projects Officer, UA Research Program 179

15 Naval Postgraduate School 191

16 Fleet Numerical Weather Central 195

17 Naval Environmental Prediction Research Facility 199

18 Civilian Manager/ Senior Scientific Analyst 203

19 Summation.. 213

20 Addendum... 219

Appendix.. 223

Glossary 227

PROLOGUE

Why did I become a naval aviator and then a meteorologist?

It all started during the Second World War when I fell in love with airplanes. I wanted to learn to fly! Then I wanted to also become an engineer, which is later refined to being an Aeronautical Engineer. This means I must graduate from college – but I have to accomplish this with minimum expense for my parents.

Since I want to earn an aeronautical engineering degree before I learn to fly, in late October or early November of 1950 I apply for the Regular Student program of the Naval Reserve Officers Training Corps (NROTC) while a freshman at Temple Junior College, in Temple, Texas; where we are presently living in staff-quarters of the VA Hospital. In the Regular Student program one earns a commission as an Ensign in the Regular Navy (USN) upon finishing the four-year program *and* graduating from college. I am to take the nation wide Navy College Aptitude Test on Saturday, 9 December 1950. I report at 0845 at the Waco High School, Waco, Texas to take this examination.

With the notification of passing this test, I am given the date, 14 February 1951, and time, 0730, for my physical at the Office of Naval Officer Procurement in Dallas. In addition, prior to this examination, I must list which of the 52 colleges/universities that have NROTC Units on their campuses I am willing to attend. Since I want to major in Aeronautical Engineering I must first determine which of these universities offer this major within their School of Engineering.

I discover the University of Texas (UT) is one of these colleges. In addition, based upon the reference information I find in the local library, its Aeronautical Engineering Department is among the top 3 in the country. Rice University in Houston has an NROTC Unit, but their School of Engineering does not include aeronautical engineering. Since

Texas A&M, at this time, does not have a NROTC unit, I never consider this school.

On my "Choice of College" card I list UT as my first choice along with a few other colleges. Following their instructions, I then apply for admission to the college of my first choice.

On Wednesday, 14 February, I have my physical exam in Dallas. This is a very thorough examination, in retrospect, I think it is one of the most extensive I ever had. All of us who pass this physical are detained for the interview phase.

This phase consists of three separate interviews by naval officers with varying backgrounds. One of these officers is a naval aviator. When he learns I want to become a naval aviator, he wants to make sure that I will stay in the program the full four years and not resign and head for Pensacola, Florida after completing just two years of college – the minimum required college to enter the Naval Aviation Cadet program. I tell him I also want to earn an aeronautical engineering degree and I will stay till I graduate.

All applicants who are still desired by the Navy have their paper work passed to a final selection committee of their State or Territory. Each of these committees, which are chosen by the Secretary of the Navy, consists of three members to select candidates that they consider "best qualified for the NROTC." One member is a prominent educator, another is a prominent civilian of the State or Territory, and the third is a senior naval officer. Each State and Territory has a quota they can approve.

A maximum of 1440 civilians and 160 enlisted personnel on active duty will be selected for enrolment in the Regular Student program of the NROTC for the fall of 1951.

By the way, even though I had to be approved by a Texas selection committee, I did not have to attend college in Texas.

In the first part of August I am thrilled when I receive notice of acceptance by UT from the Bureau of Naval Personnel. Included in the letter are my orders to report to the Professor of Naval Science (the Commanding Officer of the NROTC Unit) for the swearing in ceremony at the University in Austin, Texas as a Midshipman in the United States Naval Reserve (USNR). As such, I receive free education (books and tuition) plus a monthly pay check of $50.00.

During the first year at UT I am a Midshipman Fourth Class. At the Naval Academy new Midshipmen arrive at the beginning of summer, not at the end. They are called Plebes while they are being indoctrinated into the Navy and the Academy way of life during their first year.

The NROTC includes both "Contract" and "Regular" Student programs. The Contract Students are men who are not planning on the option of making the navy a career and will not receive any financial assistance from the navy, for their first two years. They are selected by the Professor of Naval Science among those students in attendance or who have been admitted to the college. They attend the same Navy courses at the college as the Regular Students, which is one course (and sometimes a lab with this course) each semester for four years, plus drilling (marching – inspections, etc.) once a week in uniform (furnished by the Navy). Note, depending upon your major, only some of these required Navy courses may count as an elective for your degree requirements. In addition, the Contract Students take only one cruise, which is not with the Regular Student Midshipmen. After graduation the Contract Student *may* be offered a commission in the Reserves.

In the Regular Student program you are a Midshipman, USNR and upon graduation you are commissioned an Ensign, USN (if you have not selected the Marine Corps) and will serve at least two years on active duty and four years in the inactive reserves; six years total.

The UT aeronautical engineering degree program, which is listed as nine semesters in the catalogue, plus the eight NROTC courses, requires ten semesters to complete. Regular Student program Midshipmen who have not completed their degree requirements at the end of four years, but have made normal progress within their academic field, take unpaid leave of absence from the Navy to finish their degree requirements. These requirements must be finished as quickly as possible. They are commissioned *after* they earn their degree.

In the past, Naval Academy Midshipmen and the NROTC (Regular) Midshipmen took separate cruises; but this policy was changed several years before I receive orders for my Midshipman Third Class cruise. I became a Midshipman Third Class at the end of my first academic year at UT. While on cruise we receive more money, about $68.00 per month, and of course our "room and board" is free!

As a pilot I make the transition from the "straight-deck" carrier to the "angled-deck" carrier and from being waved aboard by the Landing Signal Officer (LSO) to the use of gyro-stabilized optics to determine the proper approach path for a successful carrier landing – by day and night.

I become a meteorologist more by happenstance than by choice. I transition from manually drawn weather maps to numerically produced analyses and forecasts, and help develop the first applications of weather satellite (TIROS) data for naval meteorologists and computer applications in Project FAMOS (Fleet Applications Meteorological Observations Satellites).

Enjoy the book while learning the traditions and ways of the navy.

1

MIDSHIPMAN THIRD CLASS CRUISE

I am assigned to USS *Shannon* (DM-25) for my Midshipman Third Class cruise. Both Third Classmen and First Classmen will be together on the combatant ships of this cruise. The Third Classmen will be given more menial tasks to perform than the First Classmen.

Since the Navy will only pay for a bus ticket to Norfolk, Virginia for me to reach my ship, I take a bus from Temple, Texas to Norfolk. Other Midshipmen and I are met at the bus station in Norfolk by a Navy troop truck. The rear truck bed is covered with canvas and has bench seats down both sides. We heave our sea bags aboard and climb in for our ride to the Norfolk Naval Base.

At the base, we assemble in the open space of a large low building. We are assigned to waiting areas for our respective ships. When most if not all of the midshipmen have arrived we march to the piers, where our ships are waiting. We are all wearing our officer style navy blue uniforms with our sea bags slung over our left shoulder.

During the cruise we will wear this uniform only when departing the ship on liberty. Aboard ship we will wear enlisted type uniforms – white jumper and pants (as shown on recruiting poster, but normally without the neckerchief) or blue dungaree shirts and pants, depending upon the type of work we are performing. However, aboard ship our round white sailor hats have a blue band around the top, so we will immediately be identified as Midshipmen Third Class. First Classmen wear their combination hats (like an officer's) with a white cap cover; but the gold braid is thin, indicating that they are not officers.

Technically, Midshipmen are senior to all enlisted personnel and junior to all commissioned officers. But the Chiefs will treat the Third Classmen as enlisted men. In general, First Classmen and Third Classmen will not interact very often while on duty, except when there is

1

a Midshipman Drill of some sort, such as midshipman gun firing and ship handling. Off duty, the First Classmen will keep us Third Classmen in line.

I'm not sure which ships are in our formation, but we have two battleships, a cruiser or two, a CVL (a light aircraft carrier built upon a cruiser's hull) an oiler (no midshipmen are aboard, not a combatant ship,) and several destroyers (DDs & DMs).

DM stands for destroyer minelayer, and they were often a DD first and converted to a minelayer while under construction – such as the ship I am assigned. The USS *Shannon*, a 2200-ton *Robert H. Smith* class light minelayer, was built at Bath, Maine. She began as a destroyer (designated DD-737); and was converted to a minelayer while under construction and was commissioned as USS *Shannon* (DM-25) in September 1944. After her shakedown cruise she was sent to the Pacific in November 1944, and during 1945 she took part in the Iwo Jima and Okinawa campaigns and assisted with post-war minesweeping activities before returning to the States.

USS *Shannon* (DM-25) (US Navy Official)

This class of DM has three 5-inch / 38 caliber enclosed gun mounts instead of the four that many of the DDs built during this same time period had. She has two 5-inch mounts (51 & 52) on the bow and one (53) on the stern. Gun tubs forward of the aft 5-inch mount are quad 40-mm mounts. In addition, DDs have torpedo tubes, while DMs do not. Therefore, the arrangement of the 40-mm quad gun mounts and 20-

mm gun mounts are not the same on DDs and DMs. The DMs have mine rail tracks on the main deck down both sides of the aft portion of the ship, ending at the stern. There are cross tracks on the fantail and in the amidships main deck passageway. However, all three DMs on this cruise have their mine rails covered with metal plates and have no mines aboard.

Our ship has cleared out an enlisted compartment in the bow of the ship for the midshipmen to call "home." Both First and Third Classmen share this compartment.

I locate my assigned "bunk:" the middle rack of three, with one long metal compartment below, which is divided into three sections for all our clothes and belongings. When the racks are in the down position, one has a difficult time getting into his compartment; because there is little clearance between the bottom rack and the top of the metal compartment. The three lids are hinged at the back. Since I have the middle rack, I have the middle storage compartment. The racks are metal framed with canvas stretched and attached to the frame. There is about a 1-inch thick "mattress" on top of the canvas.

Our first requirement is to empty our sea bags and store our gear. During this action we change from our officer style uniforms to the enlisted style jumper uniforms. Our sea bags are collected and stored elsewhere in the ship, for there is no room for them in our lockers.

During the orientation lectures, which are mainly for the Third Classmen, we are assigned duty stations for watches – which rotate during the cruise – a General Quarters (GQ) or battle station – which do not rotate during the cruise – and midshipman gun firing positions – which also do not rotate. In addition we are assigned to a watch section, one of four – which generally does not change.

The loud speaker in our compartment blares, "**Reveille! Reveille! – Up all hands! – Heave out and trice-up!**" It is 0600, and our first day of the cruise has begun, soon we will be getting underway. The three of us pile out of our racks and trice-up – raise the three racks to about a 45 to 60 degree angle and set the chains to hold them in this slanted position. We put on clogs and run for the head (bathroom) – in order not to stand in line. Our head is nearby, so we shave, etc. and return to our compartment to dress before heading to the galley for breakfast.

During this preparation we hear from the load speakers: "**Sweepers, sweepers, man your brooms; give a clean sweep-down fore and aft!**"

In the enlisted cow line, which all the midshipmen must use, Watch Standers (those standing the next watch) go to the head of the line so they will not be late in relieving the off-going watch. Normally watches last four hours: 0000 – 0400 Mid-Watch, 0400 – 0800 Morning Watch, 0800 – 1200 Forenoon Watch, 1200 – 1600 Afternoon Watch, 1600 – 2000 Evening Watch and 2000 – 2400 Night Watch.

The midshipmen eat at tables in the enlisted mess. They eat together, not by doctrine, but friends want to eat with friends – or with their own kind! It seems a lot of the junior enlisted don't know what to make of us. This is the first time they have seen or been around midshipmen.

At 0745 I report to my watch station to stand my first four-hour watch. My station is in CIC (Combat Information Center – where the radar scopes are located). Since this is the first watch for all midshipmen I have no midshipman to relieve, but it is a navy custom to report 15 minutes early. This way you are briefed on the off-going person's time, rather than yours, as to the status of current conditions.

Once you relieve the other person, you are now responsible for whatever happens on your watch. If you do not wish to be responsible for the present conditions or the necessary solutions, you do not relieve the off-going watch person until the situation is resolved. (This seldom happens, and mainly occurs when the Officer of the Deck (OOD) is being relieved.)

I report to the CIC Watch Officer and he turns me over to a First Class Radarman. This sailor assigns me to an "A-scope" (no longer in use) and gives me a briefing on how to operate it and how a retuned signal (blip) will look. I have seen this type of scope in training films. Shortly it is our ship's turn to get underway.

While standing watch I see a small radar blip off our port side. I call this First Class Radarman over to see the blip; but it disappears before he arrives. The same thing happens again. The third time, he tells me to ignore it.

I think it might be a periscope of a submarine, but keep this opinion to myself; why would a submarine be at periscope depth this close to our formation? After all I don't really know what this cruise is all about and the First Class Radarman is not concerned. My intuition is based upon navy training films seen at UT.

The scuttlebutt (rumor) in the afternoon is that all the ships had been under simulated attack by one of our submarines as they were entering deep water and starting to form up in the correct steaming formation. The correct formation is with the small boys (destroyers) in a screen around the heavies (battleships, carrier, cruisers, etc.). One of the battleships, either USS *Missouri* (BB 63) or USS *Wisconsin* (BB 64), is the flagship with a Rear Admiral aboard. The flagship is at the center of the whole formation. No ship had reported seeing the periscope to the flagship! The Admiral is perturbed.

My next watch is the mid-watch (0000 till 0400). The CIC Watch Officer is chagrinned that they had not seen and reported the periscope of the submarine. I know enough to keep my mouth shut! No point in getting the First Class Radarman in trouble.

Later in the cruise, while standing watches in the communications spaces, I am given the message board to take to the Commanding Officer (CO), a Commander, in his sea cabin. I am to hand the board to him and wait to see if he wants to send any messages in response. His sea cabin is in the superstructure of the ship and I am standing on deck just outside his cabin with the hatch open. He notices that I am holding on to the side of the ship (bulkhead) between the hinges of the hatch. In a very friendly manner he says, "Don't ever hold on between the hinges! You can easily have your fingers cut-off if the hatch should close."

As you can see, this warning has always stuck with me – and yes, I have passed it on to other sailors!

When we refuel from the oiler, it is very exciting and interesting. It is another time when one has to definitely follow instructions to keep from getting injured or swept overboard, especially if something goes wrong. Those of us (Third Class) not on watch are on the main deck assisting the sailors in handling all the lines required to bring and hold the large hoses from the oiler aboard our ship. One definitely wears dungarees for this type of work.

The sea going navy is very rank conscious, except when it can't be. Refueling is one of those times when rank does not count. The oiler sets the course and speed; the ship receiving must keep station on her. Thus, an aircraft carrier may well have the highest-ranking commanding officer, but when ammunition and supplies are being passed, it is the supply ship that establishes and holds the course and speed for the process.

We do similar line handling when a person is being high-lined from one ship to the other. A Catholic chaplain is high-lined over to us for a stay of several days. To my surprise he conducts a protestant service after conducting a Catholic mass. Catholic and protestant chaplains use the same alter cross – one side is plain and the over side has Jesus on the cross.

Another time the high-line on our cruise is used is to send a midshipman from a destroyer to a battleship because he has extreme seasickness. There are individuals who can become very seasick aboard a large ship undergoing slow motions. However, much to my surprise, some of these latter people can take the more extreme motions of a smaller ship without getting sick. Who knew?

It is a lot of fun to stand watches on the bridge. If I didn't make this clear earlier, the midshipmen rotate among the different departments of the ship to stand their watches during the cruise. Three or four Third Classmen rotate between handling the helm, the engine-order-telegraph and being a lookout on the wings of the bridge. However, one does not rotate during a watch. The regular crew is also there to direct and help us.

Being on a destroyer, we are part of the "screen." The battleship with the admiral aboard has the "lead" and the rest of the ships have to keep station on the flagship. Of course there is a base course and speed. But with the wind and wave action one is constantly moving the helm back and forth to hold a heading through the water.

When you are new to this job, it can be confusing. How, you say? Assume you want the ship to go slightly to the right (starboard) so you rotate the helm slightly clockwise. The heading, according to the compass just in front of the helm, indicates the ship shifts to the left (port) at the same time. What happened? The wind and wave action more than offset the heading that would have resulted from the movement of the helm/rudder. Thus, you have rotated the helm in the right direction, but the result appears wrong. One must be confident, and apply slightly more right rudder!

After awhile one learns to anticipate the wind and wave action to hold a steadier course. As one old helmsman said, "It takes finger tips and toes to be a good helmsman!" Meaning you use a firm but light touch on the wheel and you feel the ships motions with your toes.

The engine order telegraph signals to the engine room what "speed" is commanded by the Officer of the Deck (OOD) – in terms of basic divisions of direction (forward or backward) and speed, and then by RPM. The basic divisions are: BACK FULL, BACK 2/3, BACK 1/3, STOP, AHEAD 1/3, AHEAD 2/3, AHEAD FULL, and AHEAD FLANK. Yes, full speed is less than flank speed! Then there is a little window on the side of the pedestal where one dials in the RPM commanded for each screw.

To change divisions one moves the handle(s) of the respective screw(s) of the engine order telegraph from the current position to the full forward position (clockwise), then to the full back (counterclockwise) position, then to the desired new position. In the engine room, this makes the bells ring to call attention that more than a small change in RPM is desired. There is also a sound-powered telephone connection from the bridge to the engine room – this line requires no electricity to work!

Destroyers have two screws, so the engine order telegraph has two handles, one on the starboard side and one on the port side. Thus, you can have the engines help turn the ship. The engineering people respond by moving their handles the same way in the engine room. When the pointers of the engine order telegraph that are controlled by the engine room personnel agree with the pointers set on the bridge, then the engine room personnel have correctly responded to the bridge's command. Yes, it is a command, not a request!

We have one OOD – (OODs con (directs the steering and speed) of the ship, unless the CO is on the bridge *and* relieves him) - who likes to use the engines when maneuvering the ship. The ships of the screen periodically change positions, depending upon orders from the flagship. One time I am on the engine order telegraph and following the OOD's commands I indicate different RPMs for the two screws. Several times the engine room responds by changing the RPMs but always keeping them both the same. Whoever is in the engine room does not believe the OOD wants different RPMs on the two screws. Perhaps they would have responded better if midshipmen were not involved.

When not on watch or attending lectures, the Third Classmen were often chipping paint topside and then painting the same area later or the next day. One day I am topside performing this important function when I notice the carrier, USS *Saipan* (CVL-48), has a Grumman F6F Hellcat fighter on her bow catapult (I think a CVL has only one catapult). I see the signal of the Catapult Officer for the plane to go to full power. I can just barely hear the plane's engine. Then there is no noise and people on the flight deck are running and ducking every which way. We hear later that the propeller had come off the plane just before it was to be catapulted! Sure bet that was exciting!!!!

One day while I am on the bridge standing my watch, the ships are in a refueling formation, with the oiler near the center of the formation. We are part of the screen on the forward starboard quarter. We have a radar contact of a ship steaming as if to penetrate our formation, which is a no-no any time, but absolutely is not allowed when we are refueling ships.

Our ship receives an order from the flagship to intercept said ship and drive him off! However, the order is late in coming, and there is only one way to accomplish this in the time allowed – we go to FLANK speed! We had to pass the word to the engine room, "This is no drill – put everything you have down there on line, **now**!

This is the only time I have ever been at flank speed! I have talked with few sailors who have ever been to flank speed, outside of ship trials and combat! One can feel the whole ship putting everything it has in gaining and maintaining this speed! It is fun being on the bridge, but I know it is hell in the engine and firerooms!

The CO comes to the bridge and takes command of his ship. We have trouble convincing the errant merchant ship that we mean business! There is even some talk on the bridge of manning gun mount 51 and pointing our guns at her to let the ship's captain know we are *not* going to let him penetrate our screen!

I no longer remember which country's flag the merchant ship is flying. However, we have no radio contact with him. We finally drive him off by playing "chicken" with him on a collision course. He reluctantly changes course and steadies-up on a heading that will avoid penetrating our screen.

Before arriving in English waters we enter very heavy seas! On the bridge there is a clinometer that measures the ship's roll (side-to-side) in degrees. I notice that we are routinely taking 40-degree rolls (from the vertical) in both directions, and occasionally I see a 45-degree roll. No one standing watch on the bridge seems concerned about us rolling too far – so I try to relax, while hanging on for dear life!

The regular ship's helmsman is handling the wheel. We are pitching so much in the high seas that green water is splashing against the glass across the front of the bridge! I am not talking about foam blowing off the tops of the waves. There is concern on the bridge that the glass may break! No one is permitted to be on any of the exposed decks. (This is the roughest sea I was ever in.)

In the galley/mess deck area we are served food that one can eat with one hand, like sandwiches. Under normal conditions, one learns to hold the tray by one side and keep it level while eating – but this motion is way too much to be able to accomplish this!

Also the deck is slick with spills, so one has to be very careful walking in this area – not only to not fall, but also to see that others do not fall into you. We do have a few injuries – none serious – during these conditions.

Since my commission, I have never met anyone who remembers taking such terrific rolls. Of course I have not been associating with "tin can" sailors - slang for destroyer men.

Upon arrival in the English Channel our formation is split-up into small groups to go to various ports of call. Our first port is Torquay, England. Before we arrive we are told that only some of us (Midshipmen) can go to London and we will draw straws to see who gets to go. I draw a slip of paper and find out it is "short."

We are told later that all the ones who have drawn the short slips may go to London. However, just before we arrive in port this decision is reversed – the ones with long slips may go to London. There never is a good explanation for the flip-flop.

There is a rumor going around within the NROTC midshipmen that it has been reversed because more Naval Academy Midshipmen have drawn the long slips!

In reality, I now think there was a misunderstanding of how many Midshipmen could be off the ship at one time, so the groups were reversed – but hey, at the time it made a good rumor!

Those of us staying in port are given in-port type watches, mainly sentry type (without any firearm). We also guide visitors during our "open house" periods. About half of us get to go ashore each day.

My most vivid memory is of a tour I took of the local churches with several other fellow classmen. In the bowls of one of these churches I shake hands with a man over 200 years old, a crusader! Where he is buried, his body is well preserved, with the skin looking and feeling like leather. Many of my buddies will not touch him. Yes, the guide encourages us to shake his hand. He sure has a limp shake!

We are in port about 4 days before we set sail again. It is true that you need to get your land legs again in order to walk correctly after being at sea for a while. Then you need to get your sea legs again after you depart land.

We leave port and join-up with the other ships and after about a week of doing various exercises, we head back towards England, but our group turns left and heads up the Saint George's Channel towards the Irish Sea and the port of Dublin, Ireland.

I am still assigned to the bridge watch, this time as a lookout on the starboard wing. All of the civilian Irish ships appear to be miniatures of larger ships. And they certainly sail close to us in crossing situations, the

OOD is very concerned about not hitting these little ships, but the ship with the right-of-way (us) must hold course and speed in order not to further complicate matters!

To top everything off we steam into a heavy fog. I can no longer provide bearings of shore navigation points. The ship's Navigator is also on the bridge to assist the OOD. Our group consists of all three DMs that are on the cruise.

We pick up a Dublin pilot, who is responsible for safely getting our ship into port and alongside the wharf. He is backing our ship into its slip that is in a corner of two wharfs. He has us going backwards too fast and his order for ahead full does not stop us in time! We damage one of our rudder guards. The pilot says, "I didn't realize how strong the reverse is on your ships!" Our CO puts a couple of our hard-hat divers into the water to fix the damage.

The midshipmen who did not get to go to London get to be gone from the ship about three days; so four of us Midshipman Third Class make arrangements to spend these days in a small town inn. As we get off the train an employee of the inn meets us with our transportation. It is a horse-drawn two-wheeled cart, with the passengers sitting facing outward with our luggage between us behind the driver. The ride is quaint and enjoyable. It is very interesting to talk with the Irish; their brogue is thick, but usually understandable.

The tide is really high this far north. One can walk *downward* from the main deck to the wharf when going on liberty and walk *downward* from the wharf to the boat deck (the 01 level) to arrive back aboard ship at the end of liberty. The 01 level is one deck above the main deck.

Again we are at sea in about 4 days. I am now assigned to the Engineering Department, and my first set of watches is in the engine room. This is where the turbines that drive the screws are and where the engine order telegraph signals from the bridge are displayed and answered. My job is to go around and read all the thermometers in the room. I have to crawl under and over equipment to read some of the gages. Yes, I am wearing my dungarees!

These gages are read every 30 minutes to make sure different pieces of the equipment that provide propulsion, electricity and fresh water for the ship are operating correctly. It is always hot, and when I can, I stand under a blower to "keep my cool." It doesn't help matters that we are heading towards Cuba, because this means the air temperature is increasing each day.

After about a week, I am assigned to the fireroom. The main job of the watch is to make sure that the boilers never have low water!

It is an important job, watching the water level in a small tube and adding feed water, as required, but it is boring!

As we approach Cuba we enter a naval gun firing range, and the midshipmen get to man their gun firing stations – for real firing. We have had lectures and dry runs.

Mount 51 (my gun mount) is on the main deck on the bow of the ship. Mount 52 is just behind us and up one deck (01 level). The bridge is up one more deck and just aft of mount 52. Mount 53 is also on the main deck, but on the stern. Mount 53 is not part of our Midshipmen Gun Firing Stations. These closed mounts have dual 5-inch / 38 caliber guns that fire 5-inch diameter projectiles, which weigh about 55 pounds. The caliber indicates how long the barrels are, 38 times 5 inches (the diameter) in this case.

We have a First Class Gunner's Mate who is in charge of our group of midshipmen - consisting of six Third Classmen, manning the guns, and one First Classman, acting as Mount Captain.

The Mount Captain sits high in the rear of the mount and has a hatch that opens so he can look out to see where the guns are pointing, and generally to look around – forward and a little to each side. This opening is covered with a hood that covers the overhead, the sides and the back. The whole mount rotates in accordance with an automated system controlled by the Fire Control Director. Therefore, the Trainer and Pointer positions are not filled. Without radar control the Trainer controls the left-right movement of the mount and the Pointer controls the elevation of the two guns, which move together.

Each gun crew consists of the Projectileman, the Powderman and the Hot Case Man. When signaled to load, the powderman makes sure the spade is down and then grabs the powder case – a brass shell, like the shell behind a bullet – from the powder hoist and loads it just in front of the spade. The projectileman then takes the projectile from its hoist and places it in front of the powder case and pushes the button for the gun to load. The spade rams the case and projectile home and the breech closes automatically. The gun is ready to fire.

Unlike the 16-inch guns of a battleship, the 5-inch guns do not return to a loading position! One must load them in whatever position they are in *and* as they are moving. The Fire Control Director controls each mount. Needless to say, one must be very careful not to lose a hand or foot when loading these guns.

When the guns fire, the cases come flying out the rear of each gun, and the hot case man (my position on the starboard gun) catches/guides the shell out the case-ejector chute for his gun in the rear of the mount. He wears asbestos type gloves so he is not burned by the case.

The left gun is manned on its starboard side and the starboard gun is manned on its left side. Thus, the two gun crews stand back-to-back to each other, in front of the Mount Captain.

When directed, we load our guns without any problems and then wait for them to fire, as the mount is continuously tracking our target. When the guns fire I catch/guide the case of the starboard gun out the chute with no problems. Then we all hear this god-awful high-pitched noise!

All of us midshipmen nearly panic – all at once we know the best place to be is outside the mount! Our reaction catches the Gunner's Mate First Class by surprise, and he yells at us, "Everything is fine! You are hearing the high-pressure air blowing through the barrels to clear them of all residues."

We had practiced these various gun loading positions on the practice machines at our various colleges and had seen the training film – but at no time had we ever been told of this horrific noise or heard it in the training film. My only question is: Why not?

We fire at floating target-sleds and towed targets by aircraft. We don't know how well we are doing – after all it is not us, but the Fire Control that is directing our fire.

One day, early in our live exercises, while we are shooting, all of a sudden we stop firing and I peek out the hot case chute and cannot believe my eyes! I see sailors briskly running and handling a loaded powder case from mount 52 down to the main deck and then heave it over the side.

I never hear a credible story about what happened in the mount. I think a powderman dropped the case and it got dented and no one wanted to take a chance with it. However, I do know that the CO moved the mount 52 midshipmen, then and there, to mount 53, on the fantail, away from the bridge! And midshipmen did not use mount 52 again.

Another day we are steaming in column, with our ship near the end, while the regular ship's company is firing the 40-mm and 20-mm guns at a small remote controlled plane. No ship ahead of us has hit the plane. Our 20-mm gunner that is near the bridge on the starboard side brings the plane down! He is really happy to show-up the other Gunner's on the cruise!

Our ship is ordered to drop depth charges as a demonstration. Our regular crew performs this demonstration. It is certainly impressive!

The depth settings determine the amount of water that is thrown-up into the air. The movies showing this are nearly all faked! The only way to get the big geyser of water as shown in the films is when it is set too shallow for any submarine that hasn't just performed a crash dive.

All the destroyers get to pull up to the piers in Guantanamo Bay, Cuba. The sea going Navy is very rank orientated! The ships steam in based upon the rank of the CO's – this rank is based upon one's lineal number, when required. The Chief of Naval Operations (CNO) has lineal number 1, and it is all down hill from there.

Anyway, we are tied up to the wharf when this destroyer comes zipping past us steaming way too fast. It then goes to Back Full and stops right where it should. We hear later that this CO thought he was a hotshot ship handler and liked to show-off. Of course this is hard on a ship, and if anything goes wrong mechanically one is up the creek - but good!

We are in port only a couple of days, and I never leave the ship. However, I do buy a whole pineapple while standing on the fantail from a bumboat operated by a Cuban. It tastes great! I share it with my friends.

As the ship is preparing to leave port, we are in our white jumper uniforms to "man-the-rail" during our departure, we hear the following: "Now hear this! Now hear this! Midshipman Charles Crump report to the bridge on the double! .... Belay that! .... Will Ensign Crump please report to the bridge?" Crump (not his real name) is a Midshipman who had missed taking his First Class cruise when he should have, so he has to complete this cruise before he is commissioned. For some reason he has just been commissioned in the wardroom, rather than waiting till we return to Norfolk.

The CO is *inviting* him to be on the bridge as we leave port. Just thought you would appreciate knowing the difference a commission makes!

Which reminds me of the time while enroute to Cuba, the "Plan of the Day" states a time for Midshipmen not on watch to muster for lectures; with Third Classmen on the bow and First Classmen on the stern.

Soon after we are gathered in our respective places, we hear, "Now hear this! Now hear this! All midshipmen on the bow report aft and all midshipmen on the stern report forward! On the double!" This is so the ship's officers, who have gotten it backwards, will not have to change their positions for their respective lectures.

This is a good place to relate that when a large number of personnel are going to move about the ship, such as when GQ is sounded, one MUST remember it is forward and up on the starboard side of the ship and aft and down on the port side. If a given area has only one passageway or wide ladder, then those moving forward or up are on the starboard side in single file, and all on the portside in single file are moving aft or down. Woe be to the sailor who forgets!

Our lecture on the fantail is about classified material. Unfortunately, this junior officer, a Lieutenant Junior Grade (LTJG), actually has a classified binder with him. When a page blows out of this binder and over the side of the ship, his face turns absolutely white – all his blood has drained from his face! He knows his career in the navy is over!

Then it is back to Norfolk and a bus ride home to Temple.

2

MIDSHIPMAN SECOND CLASS CRUISE

The purpose of the Second Class Cruise is to expose midshipmen to naval aviation and the Marine Corps. These are options one can pursue, after this cruise to become a Marine and after the First Class Cruise to request flight training – both NROTC and Naval Academy. All the country's Midshipmen Second Class (regular) are split roughly in half, with one half at Naval Air Station (NAS) Corpus Christi (all NROTC) while the other half is in the Norfolk area, at an amphibious base in Little Creek, Virginia, run by the Marines.

I am appointed Company Commander of the company which includes mostly our UT detachment for our first phase at Corpus Christi, and have to report a day early to help setup the facilities for the troops. One of our NROTC Instructors, Lieutenant Commander O'Malley, is also sent to work at this base for this cruise.

We all get to ride in an SNJ. This is the single-engine trainer I will later learn to fly in the Pensacola area. There is no requirement to fly and some decline the offer. The rumor is that last year an SNJ crashed and killed the pilot and midshipman.

After strapping on my parachute I climb into the rear seat and tell the pilot I am planning on becoming an aviator and for him not to hold back on any demonstration! He doesn't! I really enjoy it, and am sorry when our flight is over.

In small groups we also have an opportunity to fly in a PBM seaplane (wing span 118ft, length 80ft), which is interesting – but seaplanes are not for me. We ride out to the plane in a small powerboat. After we are all in the plane and ready to go, the pilot checks his two engines and completes his preflight check list while taxing out to the takeoff lane and turns into the wind to takeoff. We are proceeding at full power when the pilot gets the word to abort the takeoff!

A crewman has improperly closed a hatch and we are rapidly taking on water!

We takeoff after the crew has bailed out this water and the hatch is properly closed.

One by one we get to spend some time in the copilot's seat and "fly" the plane. The controls feel very heavy to me. Later I wander back into the tail to see how it looks from the gun turret. Yes, the guns are there. It is very interesting and I can imagine how a gunner might have felt while he was trying to bring his guns to bear on an enemy plane attacking from the rear! Hey, I have an imagination.

By the way, while in Corpus Christi, we are being told to be sure and take salt tablets during the day. There are small dispensers in every building and hangar.

Besides attending lectures and presentations we do calisthenics and play various competitive sports during this phase. As an incentive, the order in which the companies finish in the sports competition is the order in which they will depart Corpus Christi via Navy/Marine transport aircraft for Norfolk. This is the first year that aircraft have been used to transport midshipmen between Norfolk and Corpus Christi. Previously trains have been used.

There is a midshipman, Thomas Maggard (I think), from the University of Utah who is studying engineering. When he reads that the Marine Corps Fairchild R4Q (C-119) will be one type of the transports used, he goes to the Commanding Officer (CO) of the Midshipmen, a Commander, and informs him that at school they had accomplished a detailed evaluation of the R4Q/C-119 plane. This evaluation clearly demonstrated that this plane should have never been accepted by the military – it was too dangerous – if it has an engine failure during the climb-out after takeoff or early in the flight, it has to crash-land for it cannot maintain altitude on one engine when loaded with men/material and a large fuel load. Because of the results of this evaluation, he requests not to be assigned to this type of aircraft for the flight to Norfolk. The CO says, "No one is going to be individually assigned to any aircraft!"

The results of the sports competition are announced a few days before we are to depart. Our company comes in first, so we will be the first to leave – which means we get first choice as to quarters at the Naval Amphibious Base Little Creek, Virginia. We see that the first aircraft is a Navy Douglas R5D (C-54) Skymaster, a military version of a civilian DC-4. It is a four engine aircraft that will fly us nonstop to NAS Norfolk.

When Maggard sees the company which he is serving in – the one that is composed mostly of students from Oklahoma and Rice universities - is scheduled to fly in R4Qs; he again goes to the CO and says he will buy his own airline ticket and pay for ground transportation from the airport to Little Creek to avoid flying in this particular transport plane. The CO says he will not grant his request for civilian travel; that orders are orders, and if he does not fly on his assigned plane he will be discharged from the NROTC program!

On the Thursday morning of our departure, 16 July 1953, we pile into an R5D. After claiming a window seat, I go forward to the cockpit and am surprised to see that our pilot is a Chief. I did not know the Navy had enlisted pilots. The copilot is a commissioned junior officer. In an airplane, knowledge and experience counts more than rank 99% of the time!

I recently learned that the first enlisted pilot, called Naval Aviation Pilot (NAP), earned his wings in 1920. This NAP program was discontinued in 1947, but the pilots kept flying, mainly in support-type aircraft, until the last one retired in 1981. Some 5,000 enlisted men in the Navy, Marine Corps and Coast Guard earned their wings in this program.

Upon arrival at NAS Norfolk, we are driven by bus to the Amphibian Base in Little Creek. We choose a Quonset hut that seems to be located in the best location for our new "home." We are issued Marine fatigue type uniforms to wear, so we unpack our sea bags and load our lockers as specified by a Marine Sergeant and change into our new uniforms. Then he inspects our lockers and us! Yep, this is our Marine indoctrination.

At 2200 we are in our double-decker racks and turn off the lights as Taps is blaring over the load speakers.

However, three times during the night a person woke us, one-by-one, to ask us our name. The final time we are awakened we have to show our military identification card. At the time, no reason is given for this request, and it is conducted in whispers.

At 0400 we are awaken to Reveille played by a bugle coming from these speakers. Before we head for breakfast we assemble in a large group to hear, we think, more instructions. Instead a Marine officer tells us that one of the planes transporting our fellow midshipmen has crashed!

We are not told how many have been killed. The officer does not know what news about the accident has or will be released to the public, but the base does not want a lot of calls asking about individual midshipmen. So we are told to call our parents or other relative to inform them that we are fine.

As I tell a very sleepy sounding Dad that I am fine – he is confused why I am telling him this. He was still sleeping, and had not heard about any plane crashing before going to bed.

This is a very sobering experience for us! Who knew that expertise on the athletic field would determine who lives and who dies? In this regard, it might have been more palatable had this unit been the worst in the standing, but they were not! At lease then there would have been *some* semblance of rationality!

Over the week-end we learn that two Marine Corps "Flying Boxcars" were involved in this flight. Each of these planes carried 40 midshipmen and six crewmembers. In addition to the crewmembers, the second aircraft carried 24 midshipmen from Oklahoma University (OU), 11 from Rice University, and 5 from other schools including Utah. Maggard was aboard this plane and took a seat next to a door, so he could exit quickly. The two planes flew together to the Pensacola, Florida area and refueled at a navy airfield north of the city, named Whiting Field.

While on the ground, some of the midshipmen (without permission) switched planes.

Just after takeoff, while the second plane was climbing out from the field, the left engine failed. The pilot turned left, apparently hoping to make it back to base or over more open ground; presently they are flying over an area covered in trees. The plane *must* crash-land, as predicted by the study at the University of Utah.

The plane burst into flames upon hitting the ground and our concerned engineering-student was helped from the wreckage by another injured student, Dale Scott from OU. Unfortunately Thomas Maggard died while lying on the ground. Initially five survived – but two (the pilot, Marine Capt. C. E. Graff, and Dale Scott) would later die in a local hospital. Only three lived, a crewmember, and two midshipmen – one from OU and one from Rice.

Because a few midshipmen had switched planes, the Marines were triple checking during the night to determine who had been killed. I do not know of anyone who switched planes at Whiting Field, and I do not want to know. This accident wiped out a good portion, of the Midshipman Second Class (in the regular student program) from both OU and Rice University.

As you might expect, we did a lot of marching, physical fitness training, marching, lectures on the Marine Corps, marching, lectures on amphibious landings, marching, inspections, marching and of course we make numerous amphibious landings on the beaches. We load aboard the landing craft as they are tied up to a pier. We take turns operating the landing craft, except during the final run onto the beach.

We are told it is too easy to breech a landing craft during the beaching and there is not enough time for us to learn the fine points of the actual landing of a craft. But we do charge up the beaches as soon as the ramp goes down, with our rifles, canteens of water, etc. We practice with different sized landing craft.

Then we are sent to sea, my group is aboard Landing Ship Dock USS *Epping Forest* (LSD-4). She was made for England during WW II; but was refused by England when delivered – I think she was no longer needed. However, this ship and her sisters were made to English standards. This includes a fireroom that is pressurized. Therefore, the fireroom has double-doors. If someone opens the outer door as a person is opening the inner door – fire will immediately envelope all the ones in the fireroom and those entering! I never learned why the British had pressurized firerooms. It makes me nervous when I visit the fireroom.

A LSD is in the shape of a square "U" aft of the bridge structure. The stern is square with a large clamshell. Inside this well were landing craft, which were boarded while resting on the lower deck. Then the ship, mainly by the stern, was flooded with water to decrease the freeboard of the ship. When the clamshell was opened water came flooding in to lift all the landing craft. They started their engines and sailed out the stern. This was a faster way to get the first wave of troops ashore. We are carrying landing craft in the well.

Our sleeping quarters are those of the troops that would be hitting the beaches in the landing craft contained within the well of the ship. The racks are at least 6 deep, and maybe 8 deep. We feel like sardines! Needless to say, the only time we are in this compartment is to sleep.

The next day we are to hit the beeches again, this time from the ships at sea. We think we will be boarding our landing crafts, as they did during WW II – not so! They want us to climb down a cargo net in order to get into our crafts. So we watch, as the ship opens its clamshell and slowly floods in order to settle in the water, then the landing craft motor out the stern. They proceed to their assigned areas to circle. Upon the starting signal, one at a time the landing craft from the first circle comes along side and waits as we climb down the netting, facing the side of the ship and remembering *not* to look down. We have also been warned to make sure our chinstraps are *tight*. Sure enough, there is always that ten per cent who can't or won't follow directions! Some helmets fall, but fortunately, no one is hurt.

When my feet are nearly level with the gunnels of the landing craft, the craft swings away from the side of the ship. We have been told *not* to let our feet get below the gunnels of the landing craft, for our feet could be crushed. Likewise, if you fall between the ship and craft, you will be crushed!

I suddenly remember Dad telling me that just as he was stepping onto the landing craft at Normandy, France (during WW II) it had swung out away from the ship and he hung on for dear life with his arms, because his feet kept slipping from the wet rungs of the net.

Soon our craft swings back to the side of the ship and I drop down and scramble away from the "drop-zone." As soon as our craft is full we head back to our waiting circle. Round and round we go until the signal is given for us to head for shore.

All the landing crafts are heading towards the beech in several columns. As we near the beech, the signal is given to form in line-a-breast. When all the landing crafts are abreast of one another, the signal to increase speed is passed and we all go speeding towards the beech.

As soon as the ramps drop down, we charge up the beech to secure our assigned objectives. I am greatly surprised to see a group of individuals in bleachers, well off to the side of our landing area, observing – we had not been told that this is also a demonstration.

I am also very glad that this is an exercise, and not for real!

During one inspection the Marine Officer performing the inspection, apparently likes what he sees and says, "Do you want to be a Marine, Mister?" I reply, "No sir, I want to be an aviator." "You can be a Marine and fly." "Yes sir, but I want to be in the Navy." "Think about the Marines when you get back to college." "Yes sir!"

By the way, during our stay at Little Creek, we never have any liberty off base. In Corpus Christy we did, Jackie (my future bride) and some of her girl friends came down to Corpus and I was able to be with them one week-end evening.

The navy has flown me to Norfolk, but it is back to a bus for my return to Temple.

3

MIDSHIPMAN FIRST CLASS CRUISE

The purpose of the First Class cruise is to learn what a junior officer does aboard ship and to practice celestial navigation.

Four of us from UT decide not to take a bus; but drive a car to Norfolk, store it, and report for duty. The owner of the car lives in Houston, so we all meet there. It is a bus ride for me to Houston.

I am assigned to USS *Missouri* (BB-63) for this cruise. This is the ship that was used for the Japanese to formally sign the surrender document on 2 September 1945 – officially ending WW II. There is a plaque on the 01-deck where this ceremony took place.

USS *Missouri* (BB-63) (US Navy Official)

Fortunately, the port at Norfolk is deep enough for *Missouri* and *New Jersey* to be dockside, so we walk directly aboard. The In-Port Officer of the Deck (OOD) directs me aft to the reporting station for Midshipmen. I am assigned birthing on the first deck, which is one deck below the main deck. Compared to previous ship birthing spaces this is spacious and has a decent mattress on a firm rack. No tricing-up! The lockers are also a decent size; hey it is nice being on a battleship! In fact I am on the flagship, which means the Rear Admiral in charge of the cruise is aboard.

During the orientation lecture I am assigned Main Battery Plot for General Quarters, Mount Captain of a dual 5-inch / 38 caliber closed mount for my Midshipman Gun Firing Station, and a watch section. My first watches are in the Engineering Department, in the engine room.

My section is not on watch as we get underway on Monday, 7 June 1954, so I man the rail, as directed. As the ships are forming-up in formation, two battleships (USS *Iowa* (BB 61) and USS *Wisconsin* (BB 64), of the Iowa class, appear. Our Admiral directs the other battleship, USS *New Jersey* (BB 62), to follow us out of the formation. There were only four Iowa class battleships built for WW II; this is the first time these four ships have ever been together. As it turned out, it was also the last.

All four WWII Iowa class battleships off Norfolk, Virginia on 7 June 1954, from left to right, *New Jersey*, *Missouri*, *Wisconsin* and *Iowa*.
(US Navy Official)

Much to my surprise, the four begin maneuvering in formation, as a small group of destroyers would have. The battleships are directed to make a turn using "Standard Rudder." The respective OODs pass this command to their helmsman.

Background: each class of ship has its own standards for rudder movement. However, when ships join together on a cruise there must be only one standard that applies to all ships. If this were not true, maneuvering would be a disaster, for turns are specified in terms of a Standard.

However, only one of the OODs on the Midshipman cruise is smart enough to pass it in terms of degrees, rather than a standard, in order to avoid confusion. The other one said, "Right Standard Rudder." His helmsman used the standard for the midshipman cruise rather than for battleships!

The bridge on this class of battleship is divided into two sections, the section in front of the "wheel house" where the OOD commands and the "wheel house" where all the engine and steering controls are located. Between these two is a steel bulkhead with portholes. The OOD does not have a rudder repeater in his section to check what the rudder setting actually is. It is awhile before it becomes obvious that one of the ships is not using the same amount, degrees, of rudder movement as the other ships! Two of the ships come close to colliding – you will understand this better after I later describe my acting as OOD during Midshipman Ship Handling Drills.

It is not unusual for a ship to go to GQ (General Quarters) shortly after forming-up, to make sure everyone knows where to go during a drill and not be learning in an actual situation. Therefore, I begin asking questions on how to get to Main Battery Plot – this is where all three of the 16-inch gun turrets are controlled. I am told to go down to "Broadway" and then go forward to frame xx to find it. I discover that many decks down, in the center (side-to-side) of the ship there is a very long straight passageway that extends a good portion of the length of the ship – this is called Broadway. At each frame is a watertight door that is closed promptly after about 3 to 5 minutes of the call to GQ. After frame xx I find the hatch off broadway to Main Battery Plot.

From here I go topside and forward to find my 5-inch gun mount on the starboard side, on the 02 deck or 02 level – second deck above the main deck. The main deck is also known as the weather deck on this type of ship.

On my first engine room watch, I am amazed at how big it is – compared to a destroyer. However, over our heads are the super heated steam pipes from the four firerooms to the two engine rooms – as they are on a destroyer. If one of these parts, one is instantly killed. Yes, it has happened, but it is very rare outside combat. There is not a lot for the Midshipman First Class to do, other than observe – so I try to learn something new about each piece of equipment during each watch. Third Classmen are used to read temperature gages, as on the destroyer. It is all very interesting, but it is also hot!

Our first port of call is Lisbon, Portugal. As before, the ships on the cruise split-up, so only some of us are in the Lisbon Port. Our draft is too deep to allow us to be docked, so we must anchor a short distance off shore.

When a ship is *not* underway, the In-Port OOD is not normally one of the qualified Underway OODs that stand watches on the bridge. He is usually an officer who is in training to become an Underway OOD. The In-Port OOD stands his watch on the Quarter Deck of the ship, which is next to the officer's gangway. On a large ship, such as a battleship or aircraft carrier there are two gangways – one for officers and one for enlisted personnel. The In-Port Junior OOD stands at the enlisted gangway.

I am assigned to assist the In-Port OOD, and report on the Quarter Deck before we drop anchor and the Sea and Anchor Detail is secured. The officer, who will assume the duties of In-Port OOD when these duties are shifted from the bridge to the quarter deck, is inspecting the Side Boys. The Side Boys stand in two rows at the gangway and salute the arriving or departing high-ranking officer as the Boatswain's Mate pipes him aboard or ashore.

After he finishes his inspection he comes to me and asks, "The Mayor of Lisbon is approaching us in that launch, how many Side Boys should we have?" I reply, "I do not know, sir, we never used Side Boys on the destroyer I was on during my Third Class cruise." Then he asks, "Where would you look to determine how many Side Boys are required?" I think, wow this is getting tough. I say, "Sorry, sir, but I do not know." Expecting him to chew me out a little and then inform me where to start looking, I am surprised when he says, "I don't know either! What do you recommend?"

You may be wondering why I am reporting this, but it is a very important wake-up call for me! From then on I never assumed or trusted officers senior to me to know what they were doing – and always (almost) spoke up when I thought they were wrong or guessing.

On liberty in Lisbon with some of my buddies I see my first, and only, bull fights. When I find out that the bull will not be killed, I am disappointed. However, after seeing the first fight, I am very glad the bulls are not being killed!

In a few days we are back at sea. General Quarters (GQ), which is manning Battle Stations, is published in the Plan of the Day so people know it is a drill and not for real; and allows many people to start going to their station before it is sounded by the klaxon horns about the ship – a very distinctive "AH-OOO-GAH" sound. One day when I am at my GQ station, the muster indicates that a crewman is absent. All the watertight doors are closed, and there is no way a person may pass through any of these doors to get to his station. After about 10 to 15 minutes, we hear strange noises coming from the ventilation vents. A little later, there is a man's face at one of the vents. The crewmen pull the cover off, and our missing crewman drops into his correct station! It is much better to be late than absent.

In a few more days the formation splits-up again and heads to their respective second ports of call. We head to Cherbourg, France. The port is deep enough for our ship to tie-up at a pier, so we do not have to come and go by launch, as we did in Lisbon.

I go to Paris on buses with a large group of midshipmen. Enjoy seeing the Eiffel Tower and then seeing Paris from the observation deck, visiting du Louvre (Mona Lisa, Venus de Milo, Winged Victory, etc.) and the Versailles Royal Palace just outside of Paris. It is also fun to go to the Follies and see the elaborate settings and dances, including the Can-Can! The subways are really great, and easy to use. Yes, it is true; the French are rude and exhibit uncivilized behavior towards us, certainly not all of them, but way too many! Since we are in uniform, they know exactly who they are making feel unwelcome in their country. This behavior is a big surprise to us, having just come from Portugal where everyone tried to make us feel welcome!

After we leave port, I am assigned to Flag Plot for my watches. It is the command center for the Admiral. On this day, I am told by a member of the Admiral's Staff to monitor a PPI (Plan Position Indicator) radarscope that is on a pedestal away from the bulkheads. Our group of ships is being joined by another group of ships that includes a cruiser as command ship. There are numerous ships within the range of our radar and the Admiral is trying to make sure that the small formation of ships that are closing on us are our ships.

I place the piper on the largest radar echo, to give us a range and bearing to the cruiser, if this is our other group of ships. Whenever the Admiral approaches me, I stand aside so he can look through the hood that keeps reflections off the scope. When he steps away, I look through the hood and see the piper has been placed upon the lead ship of the formation, obviously (to me) a destroyer, by the size of the radar blip.

This switching back and forth continues – he never asks me why the piper is not on the blip he selects, and of course I never ask him why he keeps putting it on the *wrong* blip! Finally, he steps away, and I step back to my position, but he changes his mind and steps back – I accidentally step on his foot. I immediately apologize, but he never replies.

The Admiral asks the cruiser to provide her range and bearing to us. I put the piper back on the largest blip, and sure enough by taking the reciprocal of the reported bearing the blip under the piper is the other group's command ship. The Admiral peers at the radar screen again, and then has the talker report to the cruiser that he does not concur with the provided range and bearing.

I say, "Admiral, if you...." I immediately stop talking when I see his Chief-of-Staff, a Captain who is standing behind the Admiral, is shaking his head "no" at me. The Admiral does not respond to what I had started to say – probably a good thing!

When the lookouts confirm the small group of ships closing on us is ours, the Admiral has a small tantrum because these ships are in the wrong formation! Of course I wonder why he couldn't tell this by observing the ships on radar!

Another indication that **rank** does not make **right**!

Shortly after this, four of us Midshipmen First Class are invited to have dinner with the Admiral in his wardroom. The four of us get all spiffed up in our dress blue uniforms (Midshipmen do not have a formal uniform, as officers do – this is the same uniform we wear on liberty, and except for this occasion, never wear them aboard ship.) We arrive promptly at the appointed time. After the introductions, we are seated *at* a small table – I had never seen such a complete formal setting – including finger bowls. By the looks we are giving each other, I don't think any of us had ever seen this much silver, crystal and china at each place setting.

By following the Admiral's queue we all do fine. He is the perfect host! His stewards are very busy and very formal in their manner. We keep up a casual conversation during dinner. Then, to our surprise, we are invited to watch a movie, Marilyn Monroe's "Niagara." Near the end of the movie a very nice small craft goes over the falls. The only comment through the whole movie is at the end by the Admiral, when he says, "What a loss of a beautiful boat!"

I am not sure if all the Midshipmen First Class are invited during the cruise, but this is his intension.

During the latter part of the cruise our group of First Classmen has to practice what we have learned in college about celestial navigation. We have no watches to stand, but we have to do all the navigation work that is being accomplished by each Navigator aboard our ships. It is a lot of work, each morning we get up early while it is still dark and shoot the stars at first light, at noon we shoot the sun; and at last light we shoot the stars again.

"Shooting" means using a sextant to measure the elevation of a celestial body above the horizon. We use the sextant to obtain our position (latitude and longitude) by the stars/planets at dusk and by the sun at local noon. By knowing the Greenwich Mean Time of local noon – time of the highest elevation angle of the sun – we know our longitude. This corresponding elevation angle provides us our latitude. (On an aircraft the sextant has a bubble, like a level has, so it can be held horizontal for a shot; therefore a horizon is not required aboard an aircraft – at this time. No, no, a ship moves around too much at sea to try and use an aircraft sextant.)

It is a lot of work, but it is very interesting! We do this in pairs because it takes two people to properly accomplish this navigation. Each ship's Navigator uses a Chief Quartermaster for his assistant. My partner, Quint Glass, is from the Naval Academy. He does not wear glasses, but his eyes are too bad to see the stars well enough to "shoot" them. He asks me not to mention to anyone that he is not doing half of the observations with the sextant. I agree not to say anything. He tells me he is going into the Supply Corps when he graduates.

When I say mark, Quint notes the time from a watch that has been set to agree with the ship's chronometer, corrected for its known error. Then I read the elevation angle (in degrees, minutes and tenths of minutes), for him to note next to the time. We may do this two or three times for each body – then I tell him which observation I think is the best of the lot. Don't forget we are doing this aboard ship, so we do not have a level platform from which to shoot. I am glad we are on a battleship rather than a destroyer. Then we move on to the next body, which we have already looked up to see approximately where it should be.

When you cannot see your preplanned celestial bodies (because of clouds or whatever), you shoot what you can and then later figure out by the ship's heading and the relative bearing which body you were looking at.

We use the *Nautical Almanac* and mathematical tables to solve, with spherical trigonometry, the PZX triangle. Where "P" is the position of the celestial pole, "Z" is the observer's zenith (point directly over your head) and "X" is the celestial body (star or planet). From this we obtain a line of position – meaning the same elevation of the celestial body could have been measured anywhere along this line. We want to shoot three good bodies (about 120 degrees apart) to obtain three lines of position. Where these three lines cross is our position. In reality, especially aboard ship, one normally obtains a small triangle from these three lines; the center of this triangle is the best estimate of the ship's position.

The position of each ship, as determined by its Navigator, is sent to the Admiral's Staff. They offset the position of each ship by its location in the formation and then average all the positions to determine the position of the Flag Ship, three times a day. However, the secret leaks out – the position determined by the Navigator of our ship is never used, because his positions are so bad! (This may have just been a rumor – because as I recall, this is the same comment the First Classmen aboard *Shannon* (DM-25) were saying during my Third Class cruise.)

Our (Quint and my) positions are always very close to the posted position of our ship. Since we do not have a hand-held-calculator (they do not exist) there is a lot of arithmetic to do with pencil and paper – we check each other's work. We have to get the same answer since we use the same sextant and time values.

As you know, this type of navigation is rapidly becoming (or has become) a lost art!

When we come closer to Cuba the Midshipmen have ship-handling exercises. One day when I am acting OOD, I give the order, "Right Standard Rudder!" to the helmsman, a Midshipman Third Class. I notice that we are not turning, so I ask if he has heard my order; and he replies, "Rudder, right standard, sir". Still nothing happens! I am starting to wonder what has gone wrong, when I notice the bow finally starting to move to the right. Then the bow swings faster and faster! I am really surprised how fast the bow is swinging. I mention this to point out how a large, heavy ship has a lot of momentum to overcome during ship handling. It certainly handles differently than the destroyer I was on during my Third Class Cruise!

During our Midshipmen Gun Firing exercises, I have no problems being Mount Captain of mount 55. There are five 5-inch/38 caliber closed mounts on both sides of the superstructure. The five on the port side are even numbered, 50 through 58 going aft, and on the starboard side they are likewise 51 through 59. Mounts 51, 55 and 59 are on the 02 level and mounts 53 and 57 are on the 01 level. Before our first live firing I do tell the gun crews, who are Third Classmen, about the noise they will hear when the compressed air clears the barrels after each firing.

One day, I am reading a paperback book with my head outside the hatch atop the mount, while I am waiting for us to fire. Our First Class Gunner's Mate is called away to help with a problem within another 5-inch mount. When it is our turn to fire, I duck down and close the hatch above my head, but leave my book lying outside within the mount captain's hood. Since our Gunner's Mate has not returned I keep my hand near the emergency cut-off switch, in case something goes wrong. When we are through shooting, I open the hatch and find my book all torn up from the concussions of the gunfire!

Before we fire the 16-inch guns, there is a lot of work that must be accomplished away from the guns to prepare the ship for the side effects. All items are stored away; many glass areas are taped over to prevent flying glass, if the glass should break during the firing. The whole ship moves sideways through the water during broadsides from these guns! Yes, we have Midshipmen assigned to at least one battery along with the regular ship's company.

This time, I do go ashore in Guantanamo Bay, Cuba, but we cannot leave the base.

We are steaming northward to Norfolk on only two of our four boilers. The other two firerooms have been cleaned, including the bricks around the burners. Then someone lets the water level get low on one boiler, and one of the cleaned-up boilers must be brought online. The Engineering Officer is really mad! There will be no liberty for the firemen who let this happen, when we return to the states. Our ship is being readied for mothballing, and the Engineering Officer had gotten ahead of the work that had to be accomplished.

Back in Norfolk, we catch a ride to our stored car and then hit the road for Houston, via New Orleans, another interesting city. And finally, for me, a bus ride home to Temple.

4

COMMISSION - BASIC FLIGHT TRAINING

Commission, Graduation and Wedding

During my last year of college I may apply for flight training as my first assignment after commissioning. Shortly after I apply, it is off to Dallas for a flight physical and an interview. I pass both without difficulty. I am 6 feet 1 ½ inches tall and the maximum height for aviators is 6 feet 2 inches.

I complete the aeronautical engineering degree requirements after four years at the University of Texas (UT), in Austin; thanks to the two semesters of college I had in Temple Junior College. I have completed 164 semester hours of college – not all of my Junior College and NROTC courses count towards my degree. By the way, during my first semester at UT I had to take a no-credit course in Solid Geometry because McKinney High School, where Jane (my twin sister) and I attended our senior year in McKinney, Texas, did not offer this course – but it is a required prerequisite for a mandatory math course in the Aeronautical Engineering Curriculum.

The commissioning ceremony is Saturday morning, 4 June 1955, in the Gregory Gymnasium of the University. Following this program the new Ensigns of the Navy and 2nd Lieutenants of the Marine Corps walk over to the NROTC Unit (Littlefield House) and officially sign their commissioning papers and receive orders to their first duty station as an officer. We already know our first duty station, so the orders are not a surprise to anyone.

During the afternoon the Engineering Graduation activities are held inside. In the evening we receive our diplomas at the school's Graduation Ceremonies held outside near the Fountains.

Jackie and my Mother attend all three activities. Dad always worked on his sermon on Saturday!

One of the many restrictions of the NROTC program is that you will not marry before you are commissioned.

My Father marries us the next afternoon in Jackie's church in Temple.

Our 5 June 1955 wedding picture.

After the reception in the church, we drive to a café in Belton for a sendoff by our closest friends. We are driving our 1953 Chevrolet Bel Air hard-top coupe, "Brown Beauty" (dark brown body with cream color top), which we (thanks to Jackie) had bought a few months earlier.

After our honeymoon at a ranch-type resort, what else did you expect after a marriage in Texas? Yes, we ride horses.

Upon our return to Temple, we pack for our trip and hit the road the next morning for the start of my flight training in Pensacola, Florida. We go via New Orleans, a fun city.

NAS Pensacola

The Navy at this time is using Whiting Field, which is north of Pensacola, for primary flight instruction. However, before learning to fly an airplane one must complete preflight training. I report aboard Naval Air Station (NAS) Pensacola, and pass my flight physical on 17 June to begin this training. During this physical they note every scar, five in all, I have on my body; I can't help thinking this is to help identify my body – since row 39 of the medical form is labeled: "Identifying body marks, scars, tattoos."

The training consists of physical training and ground school (lectures and communications). The physical training includes a lot of calisthenics, running in sand, running an obstacle course and swimming tests – distance (on the water and underneath); treading water for a given time; floating for a given time; jumping from a high platform, swimming under water, coming to the surface with arm motions to prevent being burnt from burning oil/gasoline on the surface while getting your breath, ducking under the water and swimming, etc. till one reaches the end of the pool – fortunately, there is no fire on the water!

The most exciting is the Dilbert Dunker. The cockpit structure (no glass or metal side panels) of an airplane (SNJ) is mounted on rails. One climbs into the cockpit, straps oneself into the seat with lap and shoulder harness and waits. Suddenly the "SNJ" slides forward, crashes into the water and flips upside down while it is sinking! It is up to you to get unstraped and make it to the surface, before you drown!

However, for those that need assistance, there are two scuba divers under the water. Of course those that need assistance have to keep doing it until they do it right, use up their chances or voluntarily quit the training!

Ground school covers basic aerodynamics, meteorology, and systems – fuel, oil, hydraulic, etc. for the aircraft, North American SNJ; we will be learning to fly.

Communications training includes learning to decode Morris Code – flashing light and telegraph – at progressively higher speeds. This training is also continued at Whiting Field.

After passing all the tests, it is off to Whiting Field to actually learn to fly the single-engine; dual cockpit (tandem) trainer built by North American. This plane is designated SNJ by the Navy and AT6 by the Army Air Corps/Air Force. Not everyone who has started this preflight training advances to pilot training.

Whiting Field

After being issued flight clothing at Whiting Field we are instructed on how to bailout of the front seat of an SNJ. Then we get to practice it!

An SNJ is in "level flight," by having its tail wheel on a raised platform. The engine is running, at a low rpm, as you step up onto the right wing (not the wing that is really used to board this plane), then into the front cockpit, and strap yourself in with the shoulder harness and lap-belt. Then you advance the throttle to raise the rpm to a flying value. When signaled, you un-strap and bailout the left side (without the parachute, I believe), aiming for the middle of the left wing! You miss the wing and land in the net that is attached to the trailing edge of the wing and the front of the stabilizer on the tail of the plane.

The pilot in the rear seat reduces the power to idle, to let the next person climb aboard on the right side.

The thrust of the propeller simulates the airflow over the wing during flight. However, in the real world one aims for the leading edge of the wing – and hopes he misses the leading edge of the stabilizer during his fall. The rotation of the propeller produces a downward component to the wind aft of the left wing – that is why one always (when possible) bails out the left side of a plane, using a conventionally rotating propeller.

Yes, there is an ambulance standing by the tail of the plane, in case there is an accident!

I won't mention all the aspects of learning to fly, but first you learn to fly the plane in the air using the natural horizon (*not* the instruments), then to takeoff and finally to land the plane. Landing is a full-stall three-point (both main gear and tail wheel) touch-down in the center (left-right) of the runway; near the beginning, but not *on* the end of the runway. Our outlying practice fields are open grass fields surrounded by trees. Each flight is typically 1.2 to 1.3 hours long and we fly the 4, 5 & 6 versions of the SNJ.

Your 12th hop is a check ride by an instructor who is not your regular instructor, as a second opinion on your progress.

A check-ride is where you get an "Up" or a "Down" on the flight. The flight consists of you performing what you have learned during the last phase of your training for an instructor who has not been instructing you. An "Up" means you move on with your training, a "Down" means you repeat some of your training and get another "Check-Ride" – only so many "Downs" are allowed; before you are "washed out" of flight training.

My official Navy Flight Training picture taken aboard Whiting Field in front of an SNJ, our flight training aircraft.

Two of the procedures that you have to learn before you solo are spin and spin recovery. As I recall, on the previous hop with my instructor, I have accomplished both without problems. However, on this flight I do a full stall and purposely get the plane spinning to the left (the normal direction for a SNJ), but the control movements to stop the spin are not working – I have tried them twice if not three times before the instructor says, "I have the aircraft." When this is said, you turn loose of *everything*. He tries a couple of times, and the plane is *not* responding. Our flight rules state that if the plane is still spinning as you pass through 3,000 feet, you are to bailout. As we go through 3,000 feet the instructor does not say anything – so I sit tight. He finally gets the spin stopped, and then levels off from the resulting dive.

What do you think the instructor says? "Climb back up to 5,000 feet, and try it again." That is all he says!

I do and the plane behaves like a normal SNJ. During debrief of the flight he admits that he does not know why there had been a problem with the spin recovery. After I solo, I practice spinning and recovery on my own, which we are required to do. But I *never* forgot the time a plane did not respond to the controls during a spin! Fortunately this is the only time I ever had any trouble with recovering from a spin.

Picture of author standing next to an SNJ in California, 1994.

Your 19th hop is a check ride by another instructor to see if you are safe for solo. This flight for me is on Friday, 18 November 1955, and has gone well, as far as I can tell. I have made a few good touch-and-goes at an outlying grass field, but on final at home field I cannot get the plane set-up the way I want it before landing, so I call the tower that I am taking it around. I get back into the landing pattern and make a nice final landing.

During the flight debrief, the check pilot asks in a serious voice, "Do you think you are going to get a 'Down' for taking it around on final?" I reply, "I do not know." He then smiles, and says, "You did the right thing; if it doesn't feel right *always* take it around. By doing what you did, and thinking it might mean a "Down," I know you will not take chances in the future. This is how we want pilots to respond!" Yes, I get an "Up."

Your 20th hop is your solo – on this hop, which was delayed until the 29th because of weather, I feel confident, but also a little apprehensive. According to my logbook I made 3 touch-and-go landings at an outlying field and one final landing at home field.

I guess I should mention that during WW II the SNJ/AT6 was used for *advanced* training – the last plane you flew before you earned your wings. You then reported to a squadron that was forming up for deployment. At this time you learned to fly the plane you were going to fly in combat. At least this is my understanding.

After soloing, I flew "B," "D" and "C" designated phases, in that order; but as I remember, these were all basically different levels of acrobatics. During these phases you fly a hop with your instructor in the rear seat to learn a new maneuver or two, then go practice by yourself for a couple of flights. There was a check ride at the end of phase "B", none in phase "D" and two check rides in phase "C".

Upon returning from Christmas leave, we had gone home to Temple, they gave us two "warm-up" hops to get us back into the "grove." I am still in phase B, and my instructor is not available so I have an instructor I do not know. When airborne he asks me what I want to practice and I reply the loop. I tell him the loop was introduced to me on my last hop before Christmas leave. He says, "Fine, let's see how you do."

To do a loop in the SNJ, you must dive from level flight to increase your air speed enough to complete the loop – the SNJ is not a powerful aircraft (550 hp engine) – then you have to pull back hard on the stick to get the SNJ to make it over the top of the loop. If you don't make it over the top, the plane stalls while you are upside-down.

I do my clearing turns – to make sure there are no other airplanes in the vicinity. I push the nose over to get the required airspeed (I think it was 130 kts) then pull the stick straight back to keep my wings level through the maneuver. The main part of this maneuver that I remembered was that you really had to pull hard on the stick – so I did. However, I pulled too hard - I pull so many G's that I blackout temporarily. After I regain my senses I fake it as best I can that I had not blacked out. I do not hear anything from the instructor, so I ask, "How was that?" There is no response. So I ask again. No response. I am beginning to wonder what has happened to him when I hear this slurred voice say, "You pulled too many G's – you completely blacked me out!"

It so happens in the SNJ, that the person in the rear seat pulls more G's than the person in front. The instructors get into special positions to minimize this offset – but this time it either does not work or he has failed to be in the proper position for the loop – I do *not* ask!

My remaining practice goes well and he is nice about the incident.

Emergency spot landings are one aspect of phase "C." You never practice emergency landings without an instructor in the back seat. You will be flying along when the instructor will suddenly pull all your power off. You then select where you will try to land, and set the plane up accordingly for an into-the-wind landing. As you approach the ground the instructor tells you when to stop your simulated approach.

However, there are fields belonging to the Navy where you will actually land. At these times the instructor will tell you to land on an "emergency" field. There are white lines across each of these grass "emergency" fields. Your wheels should touch down on a white line, perpendicular to your flight path, in accordance with the wind direction! It is up to you to control your rate of decent by maneuvering the plane (no power can be used – you just "lost" your engine when the instructor pulls all the power off).

The instructor I have on this check-ride is being sort of a jerk on this hop – so I know he will pull my power when it will be most difficult to make a landing. Sure enough, it is! While I am maneuvering the plane he is suppose to be quiet (at least the others have been) to let me think; but no, he keeps telling me I am doing it wrong and I will *never* make it to the landing spot! He will not shut-up. I assume he is trying to rattle me.

I do my S-turns to lose altitude and position the plane to land into the wind, etc. My wheels touch down exactly on the white line! It is the best I have ever done on this "emergency." Does he congratulate me? No, he says, "You would never have made it if the wind had been blowing as hard as it usually does!"

It is all I can do to keep my mouth shut – but this is a check-ride, so I don't reply. My retort would have been, "If the wind were blowing differently, I would not have flown it the same way!"

I get an "Up" on this check-flight.

Then off to Saufley Field for formation flight training after my last flight on 13 March 1956.

Saufley Field

My third flight, Monday 19 March 1956, is my first day of training in formation flying at Saufley Field; I am flying the front seat of an SNJ trainer with an instructor in the rear seat. We are heading from our field to our assigned area of training. We will either wait in this area until the other airplane arrives or will see an aircraft waiting for us in the area, and join – fly formation – on him. The other plane also has a first time student (a fellow student in our flight section of four) in the front seat and an instructor in the rear. One never knows which instructor one will have before reading the flight board for the hop.

Off in the distance, about the 1 o'clock position (the 12 o'clock position is dead ahead of the aircraft), there are two tall columns of a mixture of mostly black and some white smoke. My instructor says, "I have the aircraft", and we head in the direction of the smoke. He tells me to take a good look at the smoke, because it is characteristic of burning aircraft. He also says I will never forget these characteristics.

From these characteristics and that there are two columns of smoke; he knows there has been a mid-air collision. We then head on to our training area and fly the training hop, as scheduled.

Lesson learned: A fatal accident of a fellow pilot(s) *does not* prevent you from completing your mission! This lesson is reinforced later in flight training and in operational flying.

Jackie, who is pregnant with our first child, knows I am scheduled to fly my first formation hop this day. Unknown to me, the local radio station has knowledge of this mid-air and is reporting that two training planes from Saufley Field have had a fatal mid-air collision.

While I am returning home, after this early afternoon training flight and debrief, I hear the news on the radio and wonder if Jackie has heard the story and what her reaction will be when I get home in Warrington (where we lived the whole time). This is, as I recall, the first fatal accident that has occurred in the Pensacola area since our arrival.

When I get home, Jackie is very calm about it, and yes she has heard the news, but she "knew" it did not involve me. Unfortunately, this is not the last time Jackie heard of a fatal aircraft accident and did not know if I were involved for an extended period of time!

As I recall 3 of the 4 people are killed – I think it is two instructors and one student. It is harder to successfully bailout of the rear seat than the front seat in tandem configured cockpits.

About a week later, while I am still learning two-plane formation flying, I have an instructor who does not like to let students fly anymore than is necessary. That is, with a more relaxed instructor (which I normally have), the student flies the plane from the front seat while the instructor makes comments from the rear. He only takes the controls to teach the correct method, or in the worst case, prevent a mid-air.

Thus, the student normally takes-off and lands the plane at Saufley. However, my instructor for this hop taxies the plane out to the end of the assigned runway, makes the pre-flight checks, and then takes-off. He even flies the plane to our assigned training area for this hop.

As we approach the other aircraft, he tells me, "You have it, join on the right side." As I begin maneuvering the plane into position, I momentarily feel his hand on the control stick. While flying formation for a short period of time, two more times I feel his hand on the stick, so I finally say, "If you will stay *off* the stick, I can do better."

Normally, a student would *never* say this to an instructor, but I am fed-up with him and he is not helping me by controlling the stick. He gruffly replies that he has not put his hand on the stick. I inform him, "Then we have a problem with our aircraft."

He takes control and is about to say nothing is wrong, when the stick momentarily jams. He declares an emergency, and we head back to the field.

He then begins blaming me for not performing the proper checks prior to takeoff. After I remind him, that he not I, had performed the checks, he shuts-up. He makes a wide circle with a straight-in type approach to the runway while the field is temporarily closed to other traffic.

A few days later I learn a mechanic had left a pair of pliers adrift in the right wing, and it was periodically jamming a control cable to the right aileron. I do not tell Jackie about this.

After learning two-plane formation flying and how to smoothly change positions from the right side to the left side, and vice versa, we get to fly solo, with an instructor in a chase plane. After we learn to fly four-plane formation flying with an instructor with us, we again get to fly solo with an instructor in a case plane. Four-plane formation flying means one, two or three planes are moving at a time around the leader.

Then we have our first night flight on Tuesday 27 March, with a brave instructor in the rear! We are warned **not** to get excited when we see flames shooting out our exhaust during starting and especially during the takeoff – when full power is used. In daylight one never sees these flames.

I must admit, it is a good thing that we have been warned – it would have been scary, had we not been told about this. I think it is very interesting how different it is to fly at night. We do not fly formation! We stay in a follow-the-leader mode as we fly around the field and shoot touch–and-go landings; I complete 8 of them and the final landing. I don't have any problems. The next night flight, on Thursday, we are flying solo, and doing the same thing – shooting touch-and-go landings. I complete 8, counting the final landing. Obviously, we only fly in good weather.

Then we learn to navigate by comparing what can be seen from the cockpit with the depiction shown on a WAC (World Aeronautical Chart) map. The WAC maps depict highways, railroads, mountains, lakes, parries, towns, cities, etc. That is, they are topographical maps without all the contours.

We fly in a very loose formation, we have to not hit another plane, but also keep track of where we are on our maps (yes, we have more than one map), because you never know when you are going to be called upon by the instructor to be the new leader.

After a couple of local practice hops, flying a triangle or box route in the extended local area we are off on our cross-country to a Navy field in New Orleans and return on the same day, Saturday 31 March. This is a fun flying! Both going and coming, the lead is changed fairly often between the four students. It is challenging to land at a field you have never seen before, you have no standard landmarks to help guide you.

My last flight at Saufley is on 2 April 1956

"Bloody" Barin Field

Then it is off to "Bloody" Barin – Naval Auxiliary Field Barin, near Foley, Alabama - for training in air-to-air gunnery, shallow dive-bombing, field carrier landing practice (FCLP) and finally, six carrier arrested landings.

No instructors ever fly with you at Barin. Each training flight consists of six student aviators; the first student airborne becomes the "Flight Leader" for that flight. There are one or two instructors flying separate aircraft, depending upon the mission of the day's training.

The field is called "Bloody" because of the number of accidents that have happened at this field during this phase of training. Please note, these are student aviators who have never flown in any formation larger than four planes, and off they go to fly in a six-plane formation.

Here I must digress, and explain a little about formation flying. In a two-plane formation, it is difficult for the leader to do anything that the wingman can't do. That is, it is easy (with a lot of practice) to fly a few feet away and down from the leader by keeping the same *relative* position, regardless of what the leader does – including most aerobatics. However, in this period of training *no* aerobatics are flown!

When the formation consists of four aircraft, the leader has to be aware of the consequences of his actions. In a normal four-plane right echelon (the leader is to the left with each of the other planes flying to the right and slightly lower than the plane on his left), the leader cannot make a sharp right turn (this is called turning into the echelon or formation) without making the fourth (or even the third) plane either stall or break from the formation. Likewise, a sharp left turn will leave the fourth plane falling behind because he does not have enough power to keep in position. This is why, under most circumstances, planes do not fly in left or right echelon, if there is going to be much maneuvering.

After the students become airborne they are to join-up on the student in front of them, in right echelon, and fly to the designated area for the flight's exercise of the day. Picture the first flight, six planes in right echelon being lead by a student leader with student pilots that have never flown in more than four-plane formations.

I hope you have pictured in your mind the game of, "crack the whip."

If the leader needs to turn right, when his flight is in right echelon, he signals for the formation to shift to left echelon. He then must wait the appropriate time for the shift to have been completed before turning right. This must be a very wide right turn (shallow in bank – a slow whip, if you are still thinking of the game). However, students being students, you can imagine some of the problems that occur on the first few flights at "Bloody" Barin.

In a student flight section that is several flights ahead of ours; there are two Japanese exchange officers who are *really* lousy student flight leaders. After the briefing by the instructor on the mission of the next hop, the four non-Japanese students literally run out to the flight line, barely – if at all – check the "yellow sheets", perform minimal if any preflight of the aircraft, jump in, start the engine and taxi out to the duty runway in order not to have either of the Japanese students to be the student flight leader.

"Yellow sheets" are the listing of pilot comments about problems from previous flights, how these problems were fixed, and what other maintenance has recently been performed on the aircraft. Thus these students are *jeopardizing* their lives in order to "save their lives" by not having either one of these Japanese students leading a hop.

The first missions flown are part of the air-to-air gunnery phase. A 30-caliber machinegun is mounted above the 550 hp radial engine, slightly to the right of the centerline. It fires through the propeller arc, like the synchronized guns of WW I. The butt of the gun protrudes into the cockpit and has to be manually cocked by the pilot. I don't remember the reason, but the cocking of the gun leaves the pilot's right arm black-and-blue.

Also, like WW I, there is a fixed metal gun sight to use to hit the sleeve towed by an instructor flown aircraft. The *main* objective is: Not to hit the tow plane! However, if one aims at the sleeve it will never be hit. One has to lead the sleeve, by aiming approximately halfway between the tail of the tow plane and the front of the sleeve. Yes, the tow plane had been hit occasionally; at least that is what we are told.

Another instructor flies a plane above us to observe that we have the right idea for the type of gunnery run to be flown on this mission.

Mixed in with our air-to-air gunnery flights are our shallow (20 to 30 degrees) dive-bombing flights, with little bombs, not much larger than hand grenades, which have four fins on the tails of the projectiles. These bombs are carried under both wings on small pylons.

It is during the middle of this phase our first child is born. Yes, I fly a dive-bombing hop the Saturday he is born. After seeing Jackie and our baby about 1 or 2 o'clock AM in the navy hospital on the NAS Pensacola grounds, I return home for a couple of winks sleep before driving to Barin. I do not tell any of my fellow students or the instructors that I am a new father, because I do not want to be temporarily grounded – which is standard operating procedure (SOP). It is very hard to make up a missed flight in this phase of training.

Finally we do our Field Carrier Landing Practices (FCLPs) in preparation for our carrier qualification, six arrested daylight landings on an aircraft carrier. At this time, the carrier is USS *Saipan* (CVL-48). Where: (C) means the ship is an aircraft carrier; (V) means it is to carry heavier-than-air aircraft; and (L) indicates that the hull of the ship is that of a cruiser. She had seen action during the Korean truce, and had been on my Midshipman Third Class cruise. It is a straight-deck carrier, which means the landing area of the flight deck is not angled relative to the keel of the ship.

All landings, both field and ship, are under the control of a Landing Signal Officer (LSO) using a paddle in each hand. The paddle is sort-of-like a short-handled tennis racket with horizontal strips of colored cloth in place of one-way (not crossed) strings.

Based upon the information that I found to refresh my memory, there are 16 different signals the LSO may show to a pilot in an approaching aircraft with his arms/paddles and legs. However, I only remember 13 or possibly 14 of them. I don't think the others were used by any LSO that waved me to a landing. (These LSO signals are illustrated on page 77.)

However, only two of these signals are mandatory. That is, the pilot *must* obey them. These two are the "cut" and "wave-off" signals. Of course if the pilot does not respond to the other signals the chance of getting a "cut" is slim to none.

At this time, the "cut" signal means you cut your engine (pulled the throttle back to idle), slightly but positively push forward on the stick to commence a downward motion (It is *very* bad form to dive for the deck), then flare with enough back stick pressure to do a full-stall inches above the runway/deck, in a three-point taildragger attitude. The full-stall is normal landing procedure for most Navy propeller-driven taildraggers.

Today a "cut" signal for jets means you are cleared to land on the angled deck. The attitude of jets on the final approach is that of a normal landing, the nose wheel slightly higher than the main gear. You hold your attitude right down to touch down, with no flare. Then the pilot goes to FULL power and simultaneously retracts his speed brakes! He does not retard the throttle unless the landing is arrested by an arresting wire. He immediately goes to full power, because if he waits to determine if he has caught a wire, it is too late in a jet to increase its airspeed enough to become airborne again – the plane will just dribble off the end of the angled deck into the water. Jets are much slower to respond to changes in throttle settings, compared to propeller-driven aircraft.

Then and now, a "wave-ff" means you are *not* cleared to land; you add power (and on a straight deck carrier, take the plane around the starboard side of the ship, when possible) and rejoin the landing circuit.

On my first day of FCLPs our flight flies from Barin to an auxiliary outlying field and immediately enters the landing pattern. This is our first time to fly the SNJ at 60 kts in level flight – a speed that is *just* above stalling speed. One has to adjust the throttle slightly to compensate for the difference in lift between flying over a plowed field and flying over a grove of trees; both of which are within our landing pattern. You really have to become intimately acquainted with the particular plane you are flying, because of the low speed being flown, and no two planes fly exactly alike.

In a left-handed landing pattern, the only pattern used around Navy fields, the 180-degree position is where your left wing tip is pointed directly at the beginning of the runway (stern of an aircraft carrier). From this position you commence your landing approach. Prior to reaching this point you are flying level, have gotten your plane into a "dirty condition" (wheels down, flaps down and for arrested landings, tailhook down) and have slowed to your desired initial approach speed.

As you commence your left hand turn, you pull off just enough power to compensate for the airspeed that would otherwise be gained by the loss of altitude as you descend to the runway (carrier deck). You pick-up the signals of the LSO when you are about 45 degrees off the final heading of the runway (deck), and respond accordingly until you get the "cut" or "wave-off" signal.

The (~1952) picture is of an SNJ receiving a "wave-off" from the LSO during FCLP at an outlying field of "Bloody" Barin. (US Navy)

On my first approach I fly a good pattern and receive a "cut" signal. I make sure that I do **not** dive for the deck as I transfer my attention to the runway ahead. Out of the corner of my left eye I immediately see the LSO signal a "wave-off" – a mandatory signal. But that's impossible, isn't it – a "wave-off" *after* a "cut?"

I am confused. I add power to wave-off, but then think the signal might be for the plane behind me, so I pull the power off again. But I reason the last mandatory signal was a "wave-ff;" so I add power; but the first mandatory signal may be the one I should obey; so I pull the power off again – when my right wing stalls I'm in *big* trouble!

The plane immediately falls off to the right. I stand on the left rudder, to bring up my right wing and stop the right turn, while adding power – which also helps raise the right wing. Under these conditions one must never ram on full power, because the plane will probably torque roll. (Many high-powered propeller-driven planes have torque rolled into the water during a wave-off and following a bolter on an angled-deck carrier. A torque roll is when the torque of the engine is not countered by the air flow over the wings to keep the plane from rolling in the opposite direction from the propeller. Thus a torque roll is always to the left in conventionally powered propeller-driven aircraft. (Sequential pictures of an AD in a torque roll are shown on page 124.)

I end up flying about 45 degrees to the right of the runway with my wheels just off the surface and heading straight towards a row of SNJs waiting to be flown by another section of students. These student pilots come running out of the ready shack to watch me crash when they hear the crash truck's siren. This truck is racing directly towards me with its red lights flashing. It is a very interesting sight!

Having regained control of my plane, I raise the gear and just miss hitting these parked planes as I climb and then turn left to rejoin the traffic pattern.

I conclude that I had not shown enough downward motion of my plane after the "cut," so I apply slightly more forward pressure after my subsequent "cuts" – six more this flight. After each landing one immediately adds power, takes-off and rejoins the landing pattern.

I am told during the flight debriefing at Barin that he (the LSO) had given me a "wave-off" because he thought I was going to stall and crash on the runway. He had not seen enough downward motion after the "cut." This confirms what I had already concluded.

By the way, even though I have never met another naval aviator who has ever heard of a pilot receiving a "wave-off" *after* receiving a "cut;" I doubt I'm the only aviator to ever have this happen to him – but, it is certainly a rare event in naval aviation.

I have a total of 81 first pilot flight hours. Our section of students and other sections are qualifying off the coast of Pensacola, Florida on Thursday, 7 June 1956. At this time, no pilot may earn his "Wings of Gold" without becoming carrier qualified during Basic Flight Training prior to entering Advanced Flight Training in an operational-type of craft – attack, fighter or multiengine (land or sea) aircraft, helicopter, or blimp (lighter-than-air).

USS *Langley* (CV-1) and USS *Saipan* (CVL-48) (US Navy Official)

The USS *Jupiter* (Collier #3) was converted into the Navy's first carrier, the USS *Langley* (CV-1). This conversion was completed in March 1922. She had a flight deck of 534 feet by 64 feet. The USS *Saipan* (CVL-48) was built upon a cruiser hull and was commissioned on 14 July 1946. She has an overall length of 648 feet and a beam of 76 feet.

The vivid memory I have of the actual carrier qualifications is how small the carrier, USS *Saipan* (CVL-48), looks as we fly the "Dog" pattern while waiting for our "Charlie" signal – to descend and join the carrier landing circuit

Included in the slight modifications of an SNJ for carrier landings is the addition of a tailhook; the operation of which is all manual. The pilot pulls a T-handle on the left side of the cockpit to extract a pin through the shaft of the tailhook, which allows the hook-end to swing to the down position, thanks to gravity. A flight-deck crewman has to swing the tailhook up and snap it into its "retracted" position, after each arrested landing. On a straight-deck carrier, you are either stopped with an arresting wire or with the barricade. There is no bolter" (touch-and-go landing) as there is on more modern angled-deck carriers.

I have made 3 arrested landings, with no "wave-offs," and am glad all is going well. Then, while approaching the stern of the ship, the LSO signals with his paddles that my tailhook is not down, followed by the "wave-off" signal. I add power and take it around the pattern again. There is no indication in the SNJ cockpit to inform the pilot of the position of the hook. Abeam the ship I really pull hard on the T-handle before I commence my approach – same result - no hook and a "wave-off." On the radio, the carrier's Air Boss orders me to fly to NAF Barin and return with another plane.

Back at "Bloody" Barin I take the time to run into the head (restroom) to relieve myself, before heading back to the aircraft line to check the plane that has been assigned.

As I open the outside door I see an SNJ making a wheels-up approach. The plane "lands" beautifully on its belly and slides along the runway until it stops. Out pops the pilot, unhurt. It was planned, not a mistake by the pilot. He could not get his wheels down.

SNJ Qualifying Aboard USS *Monterey* (CVL-26) (Don Chin via author)

After returning to the ship I complete my remaining 3 arrested landings. Once airborne following my last required landing, I am anticipating orders to join a formation of other students to fly back to Barin; so I am disappointed when given the signal "Dog." This means I am to keep circling the ship above the "Charlie" pattern.

After an extended period of time I am given a new signal - surprisingly it is "Charlie." So down I go to join those still qualifying. After I land and my hook is cleared of the arresting cable and manually raised to its up position, sequentially stationed aircraft handlers along the flight deck signal me to taxi – and taxi some more! I follow their signals and end up near the bow – certainly not enough deck left to stop, go to full power, release brakes and takeoff from the deck. This is how we are getting back into the air after each landing, known as a deck launch. The next thing I know my heart is in my throat, because the whole deck under my plane has given way!

Unperceived by me, they had spotted my plane on the forward, center-line flight deck elevator and suddenly, without warning, I am dropping to the hangar deck. I taxi where directed, and secure my engine when signaled. Then all the men who have secured my plane to the deck quickly walk away.

Now what?

I climb out of the front cockpit and wait next to my plane. This is the first for me to stand on a carrier. No one comes to explain why I'm on the hangar deck. I finally see a passing crewman on the far side of the hangar bay and I call, "What is going on?" He happily replies, "The ship is on its way to New Orleans for some liberty and fun!" And here I stand in my flight suit with little money in my billfold – some fun!

Eventually, a LSO approaches me and says, "Get in the back; we are going to the beach." I have not met him before, and he does not introduce himself. Since I am now a passenger, it is interesting to be able to observe more of what is happening on the flight deck of the carrier prior to our launch. I am relaxed and enjoy the deck launch and flight back to Barin field. The pilot remains quiet on the flight and I ask no questions.

So I never learn why he had selected me to make an extra landing, or my plane to fly to Barin.

Corry Field

Then it is on to Corry Field for instrument flight training. This will be the first time a student flies the SNJ from the rear seat, with the instructor in the front seat. But first, one must learn to "fly" the Link Trainers – enclosed blue boxes, with little wings and tail that pivot about a point directly below the "pilot" of the trainer. They bank (rotate laterally) and pitch (nose up and down) at the same time. Their travel is limited, but more than enough to simulate normal instrument flying. (A Link Trainer is pictured in a museum on page 155.)

One learns to fly specific patterns, involving half circle, full circle and straight sections at a specified altitude and with given time periods for each section. A standard rate turn is 3 degrees per second – which means it takes 2 minutes to turn through 360 degrees, back to your original heading (regardless of your airspeed). Some of the patterns are flown at half standard rate, and others are a combination of the two. With the link, all "flying" is in no wind conditions.

Then we must learn to fly a real plane on instruments – which I think is easier than the link trainer – from the rear seat of the SNJ. When instructed, you pull a cloth hood, which is attached behind your back, over your head and fasten it to the cockpit frame above and at both sides of the instrument panel. This prevents you from seeing outside the cockpit. The hoods are either white or black. I always preferred the white hoods.

If one tries to peek *and* fly on instruments it is very easy to really get screwed-up! I think trying this is what causes most of the problems with students that are having trouble. They try to cheat, and this just makes it worse for them.

I find it challenging, but enjoyable – I quickly learn to believe the instruments and relate to them. However, many student pilots talk about how they really dislike this phase.

Once one can properly fly a plane on instruments, one learns a little instrument flying. That is, to navigate from point "A" to point "B" by staying on an airways low frequency radio beam and make simulated instrument approaches to fields – never landing.

Once this training is finished on Monday, 25 June 1956, I report back to NAS Pensacola for my flight physical and orders. To my big surprise, I fail the depth perception part of the eye exam! I am very upset! I'm told to go home, rest and return the next day to retake the eye exam. I do, and this time I pass. My orders to Advanced Flight Training arrive by the end of the month – I have chosen Propeller Attack, because I want dive-bombing training and this is the only way to get this training. As a child during WW II, I vacillated between wanting to be a pilot and a bombardier. This way I can be both at the same time!

I want to transition to jets after my first squadron tour in operational propeller-driven attack (Skyraider) aircraft.

5
ADVANCED FLIGHT TRAINING

Instrument Training

I get my first choice, Propeller Attack, and report aboard Naval Air Station (NAS) Corpus Christi on Thursday, 12 July 1956. (This is where I had started my Midshipman Second Class cruise.) Here we are introduced to the pressure chamber. We are issued oxygen masks for the first time, and it is time to learn to use them properly. By decreasing the pressure in the altitude chamber we are taken to higher and higher "altitudes." One at a time we take our masks off while performing different simple tasks to demonstrate that you do not know that you are having problems with a loss of oxygen, until it is too late!

We are to ware an oxygen mask on any flight above 10,000 feet during the day and 5,000 feet during the night. The oxygen at night at a lower altitude is to improve night vision.

Then our section reports aboard NAS Auxiliary Cabinass Field for advanced flight training. The first phase of this training is instrument flying in the North American T-28B; shown on the next page. Note instrument hood in the rear.

Again, I really enjoy flying on instruments – in the back seat. On these flights we always fly with our oxygen masks on because we usually fly higher than 10,000 feet. My first flight in a T-28B is on Tuesday, 31 July 1956.

One requirement of the school is to learn how to plan and execute a flight in accordance with Instrument Flight Requirements (IFR) and then file and fly an IFR cross-country flight. Knowing I need this type of flight, ENS Powell, a flight instructor at the school, asks if I would like to take the required flight with him. Since he is going to visit his parents in Tennessee, he suggests that I spend the night at his parent's house. We will leave the coming Friday night, the 24th of August, and return Saturday night. After I agree, he explains the flight will be to the Tullahoma Regional Airport in Tennessee via the Strategic Air Command (SAC) Air Force base outside Lake Charles, Louisiana; both going and returning; so I can do all the required airways flight planning.

North American T28B (National Museum of Naval Aviation)

After flying a 2.0-hour instrument training flight with my regular instructor, LT Budd, I meet with Powell in the late afternoon to show him my work. He approves, so off we go to the weather office to get our weather brief and have this information added to our flight plan, DD 175. This is my first experience of seeing a meteorologist in action.

Just after the night takeoff he turns the plane over to me to fly and to speak with air traffic control (ATC) in accordance with our flight clearance. I have no problems flying the airways and making the required reports to ATC. At the conclusion of my instrument approach to the Lake Charles Strategic Air Command (SAC) base Powell takes control and lands on the runway directly in front of the plane. Fortunately for me, before I commenced the approach Powell briefly explains what will happen when we land at this Air Force base; but I am still not totally prepared.

Before departing the duty runway we have two or three jeeps, on each side of us, escorting our plane with mounted machine guns pointed at us. Following directions over the radio we taxi to the refueling area and must remain in our plane during the refueling – which is not allowed at a Navy field! The Air Force guards are taking no chances that we are not in a bogus Navy plane sent by the Air Force to test their procedures of preparedness for a surprise attack.

I am not scared, but it is unnerving to have these guns manned and always pointed directly at us. There is no joking or extraneous comment by these guards. After our plane is refueled and we have our air traffic control clearance for the next leg of our flight we follow the towers directions to the duty runway – with the jeeps still in escort!

When we approach the Tullahoma Regional Airport, Powell takes control of the plane and calls the tower to have the runway lights turned on so we can land. After Powell taxies the plane to our tie-down location and we climb down to the ground, we get to tie down our plane. We have flown 4.7 hours.

Needless to say, at Powell's parent's house I fall right to sleep when my head hits the pillow, this is the most hours, 6.7, I had ever flown in one day.

We leave after nightfall the next evening. The runway lights are promptly turned off after our departure. The flight to the SAC base goes well and I make another instrument approach to the field. And yes, the armed guards in the jeeps take the same actions as before.

When we are about 3/4ths of the way to Corpus Christi, Powell asks a simple question, "Why do you have to keep retarding the throttle to maintain a constant airspeed?" I hadn't given any thought as to why I had to do this; one just does what one has to do!

When I reply that I do not know. He asks me, "What about the weight of the aircraft?" I would like to say that I immediately responded correctly, but I didn't. At this stage of my flying I had never thought about the weight of the fuel as being significant in comparison to the total weight of the aircraft. And that as one flies, one burns fuel, and the plane progressively gets lighter – and over time, light enough to alter the power requirement. Fuel weighs about 6 lbs per gallon.

At Cabinass field I make another instrument approach with Powell landing the plane. All of these instrument approaches are "simulated" – meaning the weather was not actually IFR. One does not get credit for "true" instrument flying unless the weather is IFR at the time. Of course under the hood I could not have told the difference!

Near the end of my instrument training my usual instructor, Budd, has me make a takeoff on instruments, he makes no corrections – this is not part of the curriculum. He aligns the aircraft with the runway before turning the plane over to me.

At the completion of the final instrument check flight, flown with an instructor who is not your regular instructor, the check pilot swaps places in the plane with the student, if he has passed. This is so the student, who has been flying under the instrument hood, can get a good introduction to the local area, land marks, etc.; because from now on the student will be flying alone in a Douglas Skyraider.

By this time in the Pensacola area, a student is learning to fly starting with the T-34 and then progressing to the T-28. Some of the students behind us have transitioned from the SNJ to the T-28. My group of students is just about the last to have flown only the SNJ. I say this to explain the following.

I fly this check flight on Tuesday, 4 September 1956, qualifying me for a standard instrument rating. After we change places in the airplane, I am looking over the instruments from the front seat when my check pilot says, "Well, start the engine and let's get going." When I inform him that I do not know how to start the engine because I have never flown the T-28 from the front seat, it really surprises him – this shows how well I have flown the plane on instruments!

After the area orientation flight, my check pilot lets me shoot 2 touch-and-go landings before a final landing at Cabinass field. This is my first time to land with a tricycle-landing-gear. One touches down on the two main landing gears and then eases the nose wheel to the runway. It's nice!

A Chief who keeps the records of the instrument school tells me while I am checking out that he was looking forward to meeting me. His search of the records has revealed that I have made the highest marks ever given to a student by this school since they had started using the T-28 a few years ago.

I told you I like flying on instruments and the T-28 really handles well!

Now it is on to learn to fly the Douglas Skyraider, AD-1, at the same field. The first model, AD-1, of the Skyraider was built in 1946. Initial pre-production versions of this dive-bomber were not named Skyraider, and were flying before the end of WW II, but never became operational.

Our flight instructors are flying the AD-4. It is rumored that students are more expendable than instructors (which is true!), and no instructor will fly the AD-1.

After an in-depth explanation of every system of the plane over a period of about two weeks, it is time to fly – well, almost. Our first requirement is, when cleared by the tower, to line-up on the duty runway, increase the throttle to full takeoff power and get the tail up enough to have the plane level; then pull the power off, get the tail wheel back on the deck and go to the end of the runway before pulling off onto a taxiway. The object is to keep the plane on the runway during these large power changes. This requires a lot of proper rudder control!

With the power available in this big engine (2,500 hp) everything happens very quickly. It is a big challenge, but very thrilling! Then it is back to the other end of the runway for my first flight in the Skyraider on Tuesday, 25 September 1956.

We had been warned to get our canopy closed before reaching 130 kts. I'm not sure of the exact speed, but it is close to this value. We all laugh and think this will be no problem.

WOW! It is very difficult to accomplish this – everything happens so fast on this first flight.

Our AD flight training section
Our Instructor, First Lt. Sherman (Marine Corps), is in the center;
I am the tallest one. (US Navy)

After we finish our four familiarization-flights, we must follow tradition: which is to go to the Officers Club on the Friday evening of the following week to "kill" a bomb of beer – the nose of a bomb casing has been removed, the body filled to the brim with beer, and set on its tail fins - it is very bad luck not to finish all the beer! Drinking is not allowed 24 hours before a flight, so Friday and Saturday evenings are for partying.

I mention this tradition to Jackie sometime during the week of our familiarization flights. However we both forget it.

Then we practice formation flying, including some mild acrobatics! Our instructor is very pleased with our abilities to hold formation and change positions. He has us fly in trail – each plane directly behind the plane in front – while he does all sorts of gyrations to see if we can follow his lead and stay together. We do well.

Friday morning of the next week I tell Jackie we will be going to the dive-bombing range today – our first attempt at learning true dive-bombing (dive angels 60 to 80 degrees - at "Bloody" Barin our dive angles in the SNJ were only 20 to 30 degrees). I do not remind her about my going to the Officer's Club, because I have forgotten about it.

Early in the afternoon, during our flight brief for the dive-bombing range, we are informed that a student has "gone-in" (fatal-crash) on the range next to the range we will be using – for us to ignore whatever we see, including any smoke, etc. - that we must *concentrate* on what we are doing – or we will "go-in" too!

Reinforcement that death of a fellow flyer does not prevent you from flying your mission!

I could not resist taking a quick peek as we fly by – I see the burned spot where the plane and pilot perished, aircraft wreckage and several vehicles.

The dive-bomber versions of the Skyraider have 3 solid dive brakes - one each side of the plane aft of the wing and another one on the bottom of the plane, also aft of the wing. There is an option of deploying only the bottom brake, as a speed brake. The dive brakes slow the diving speed to provide more time "on-target" to improve one's aiming, before it is time to pull out.

However, the side dive brakes disturb the airflow over the horizontal tail surfaces – stabilizer and elevator (front-to-back). To counteract this, the pilot controls the angle-of-attack of the stabilizer. Part of your pre-flight checklist is to properly set this angle for normal flight. Prior to extending the three dive brakes, before rolling in on the dive, the pilot *must* decrease the angle of attack of the stabilizer to its proper dive setting – depending upon dive angle and bomb load. If this adjustment is not countered by forward stick pressure, the nose will rise – experienced pilots can make this adjustment while keeping the plane in level flight. If the pilot forgets this trim adjustment of the stabilizer – he will not be able to pull out of his dive!

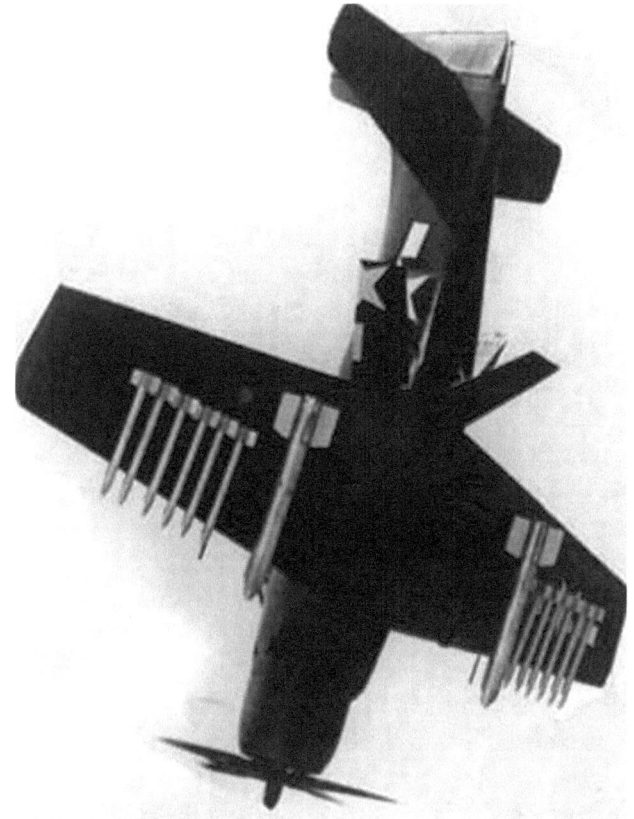

AD-1 in Dive with all three Dive Brakes Extended (US Navy Official)

Aviation Historic Note: The Skyraider had what has become know as the *flying tail* before this innovation was determined to be needed on supersonic aircraft.

I really enjoy the mission – it is thrilling to go hurtling towards the earth as you line-up your sight, allowing for the wind, to hit the bull's eye of the target marked on the ground. The altimeter is unwinding at a terrific rate – but you *must* remember that it is lagging your true height – you are going down faster than your altimeter can "unwind." You must keep your plane in balanced flight in order not to "throw" the bomb off-target. Of course the first time everything seems to happen so much faster than on later flights!

Our instructor flies in a circle around the target at 1,000 feet, we are *not* to fly lower than his altitude. We are also to pull less than 6-Gs during our pullout. We do not wear g-suits to compensate for the Gs.

Remember 1-G (force of gravity) is what one experiences while walking around. If one weighs 180 pounds and pulls 5-Gs on the pullout, one weighs 900 pounds during the pullout. Because we are not wearing G-suits, we gray-out a little on these pullouts,

Once you release one of your bombs you pull back hard on your stick, once your nose is level you retract the dive brakes to improve your rate of climb back to diving altitude – around 6,000 to 8,000 feet.

Because of the lift of the plane's wing the angle of the plane is greater than the dive angle. With the 80-degree dives – which we fly later - we have junk floating around in the cockpit because we are slightly inverted during the dive!

When our first dive-bombing hop is over, and we have been debriefed, I think I will be going home. But I am reminded that it is time to celebrate finishing our familiarization-flights! So with our instructor, we go to the O-club to "kill" a bomb of bear – to bring us good luck. Our instructor can't help us with the bomb of beer.

At this time I normally did not drink alcohol, so I tell my fellow students if I am last I'll do what I can for our section – 6 students. They don't trust me to be last, so I am fifth. I drink all I can (it must be ONE drink from the bomb), and then pass it to our last student to finish the bomb. He does, and complains I have not left him enough.

Yes, we students have finished it – so we are going to have *Good Luck!*

The time sort of gets away from me, and I do not head home until after 2100 – no, I have not called Jackie; but then neither has any of my fellow airmen called their wives.

Unknown to us students, the death at the bombing range has been on the radio – several times during the afternoon. The following is how I remember what Jackie told me happened, but Jackie does not remember it exactly this way.

Jackie hears the news of a fatal plane crash on the Cabinass dive-bombing range, but assumes it is not I. When it is time for me to come home, she begins waiting for me in front of our apartment with our son and Boxer dog; as she usually did. I do not show up in a reasonable time.

She talks with a neighbor, the wife of another student pilot who is not stationed at Cabinass Field. But the neighbor has not heard any more than what has been on the radio.

So Jackie goes back to our apartment and waits awhile longer outside, then gives up and goes inside to wait. She tries to remember what I told her in Pensacola about when the chaplain would come to notify her in case of my death. So, until it gets dark, she periodically looks out the window to see if a Navy car has arrived in our apartment parking area.

When I arrive home, I think she will be a little mad because I am so late – but I am not expecting her to be so *mad* and *relieved* at the same time!

On one dive-bombing hop while flying to our target, I suddenly feel a continuous squirt of liquid hitting me on my neck. I wipe with my gloved hand, and see by its color that it is hydraulic fluid. This means my canopy operating system has developed a small leak – but the pressure of the hydraulic fluid makes it squirt.

So I turn the master hydraulic valve to "OFF." No more squirting – but I need hydraulic pressure to open (and keep open) my three dive brakes during each dive-bombing run.

Just prior to "rolling-in" on the target (we carried about 8 bombs – we dropped one at a time) I put the master hydraulic switch back to "ON" and pull it "OFF" after I closed my dive brakes when pulling-up off the target.

Of course I have to turn the master switch back on to lower my flaps and landing gear, and to have power boost for my brakes during the landing. Thus I must bring my flight suite home to be washed – the mechanics tell me that hydraulic fluid is *not* good for your skin, etc.

Another time, my gun-sight (also bomb-sight) illumination goes out about halfway through our bombing runs. So, I pull out my trusty grease pencil from the shoulder of my flight suite and mark my windscreen where I estimate the pip has been – anything *not* to abort a hop!

During instrument flight training in the AD-1, you pull a hood over your head, which prevents you from seeing outside. You only have your flight instruments to inform you of what your plane is doing. You must train yourself not to use your ears or any other senses to imply what your aircraft is doing! One of the exercises is to do a full stall and recovery, while *preventing* the plane from spinning. (I never entered a spin during these exercises.)

When one is flying "under the hood" in a single seat aircraft there is always a chase plane to ensure safety.

On one occasion when I am under the hood there were two chase planes - our "extra" instructor and a fellow student of our section. This extra instructor had recently joined our flight to learn from our instructor how to teach a flight of students. Remember, the instructors fly the AD-4 while we students fly the AD-1.

When requested, I gently raise the nose and simultaneously reduce my power to stall the aircraft. Just before my plane stalls, I hear a gasp on the radio and "know" something bad has *almost* happened.

Back on the ground I find out from the other student that the instructor is flying on my tail (which is *not* where a chase plane is suppose to fly) when he tells me to do a full stall. My plane apparently slows down faster than his plane, because he nearly rams my tail. His propeller would have chewed off my rudder – at least.

Later in our training we begin night-formation training. One night, I have a rough-running engine not long into the flight. So I inform our instructor of my problem and that I am returning to the field, as a precaution.

I get another plane and re-join the formation. A little later I see that this plane's hydraulic pressure has gone to zero. I inform our instructor of my problem. Without hydraulic pressure I have no landing gear, flaps and no power-boost on the brakes.

Our instructor joins on me and asks me to reduce speed below the maximum allowed for lowering the wheels, and then place my landing gear handle into the "Down" position. My gear does a "free-fall," but does not lock down, based upon my landing gear indicator windows.

So I dive and pull sharply up to pull Gs to force the gear to lock down. After several attempts, I have success. Our instructor looks me over the best he can, using the lights from the city, and tells me to return to the field.

When your flaps are up, you must land faster than when your flaps are down. You need the same amount of lift, so without flaps the airflow over the wing must be greater to produce the same amount of lift.

When landing faster, you need your brakes more – right? It takes most of the runway before I get the plane stopped as I alternately stand on the brakes, left, right, left, right, etc. If you use both brakes at once you nose over! Without power-boost on the brakes it is not safe to taxi, so my plane is towed back to the line while I sit in the cockpit.

Naturally I do not tell Jackie about this! Unfortunately, a few days later one of my fellow students happens to mention to Jackie how cool I was when I had had two emergencies on one night hop. There went my credibility!

Our ground attack flights are a lot of fun, we get to strafe with our two 20mm wing mounted cannons and fire rockets at shipwrecks on a navy section of the beach; which is near our bombing targets. Later armed versions of the Skyraider have four 20mm cannons.

On Friday, 9 November, we get to fly with a drop tank on a round robin WAC (World Aeronautical Chart) map navigation hop. On the ground, during the pre-flight of the plane you must unscrew the cap and check that the auxiliary fuel tank is full, because there is no gas gauge for this tank in the cockpit. After you start the engine you also switch to the auxiliary tank to ensure that you can draw fuel from it. Once airborne and established on your outbound heading, you switch to this tank and either time the fuel drain, or fly until the engine coughs – the latter being the recommended procedure. As indicated in the Basic Flight Training section, a WAC map indicates towns, highways, railroads, major power lines, rivers, mountains, and etc. for visual ground navigation. I log 4.0 hours on this flight.

The next day it is another WAC map navigation flight, but this time it is to New Orleans! We are each carrying a drop tank, so we fly almost to Dallas before we turn towards New Orleans in order to prolong the flight. As you may recall, my Saufley flight section had flown to this Navy field during Basic Flight Training. We land on the same runway, and the runway really looks short this time! I log 4.3 hours on this trip. We fly back on the 12th, without the drop tank being filled, so we take a fairly direct course, but still using WAC maps, and I only log 3.1 hours.

We also have one low-level navigation hop to our "target." Not sure what arrangements have been made with the ranchers in this area, but we are flying lees than 100 feet above the ground with a high power run-in to the target on the last leg. This is dead reckoning navigation, based upon the wind and a plotting board on your lap over a designated zigzag path. This is a one-on-one hop, one instructor for each student; so only two planes fly at a time. There are no landmarks to indicate when to turn (at least not for the student); it is strictly time, heading and ground speed.

This instructor is perturbed at me for using my head before we man our planes. I had gone to the weather office and received the forecast winds for the area of the exercise, and had worked out all the headings and speeds before our flight – which is what I thought any good aviator should have done.

His idea is that just the basic heading and distances should have been known, then we would fly over the ocean and estimate the wind offshore from the waves and assume this would be the wind over the exercise area. While flying at low level I'd determine the headings and air speeds for each leg of the mission. He said I'd have to redo all my work after we flew over the ocean, which I agreed I'd have to do – knowing full well I was not going to do any such thing!

Once over the starting point, which the student can recognize, the clock starts ticking!

I hit the target dead on (location and time) – we are flying so low I could not see the target until we are almost on top of it; I didn't have to make any last minute corrections.

While taxiing back to the line, I make a lot of grease marks on my plotting board as though I had made changes in the air, in case he looked at my board.

All he says to me during the debrief is, "You did OK." Of course an "OK" is fine with me, since it is the best grade one can earn from the LSO on a carrier landing!

Guess it is time to tell you about the time I get hurt!

At this time single seat carrier aircraft are always boarded from the right side. This side is closest to the island of an aircraft carrier, where the pilots step onto the flight deck. There are three hand-holds, starting in the top right side of the fuselage insignia. A step is provided in the side of the fuselage below the opening of the cockpit.

The AD-1s burn a lot of oil – in fact the mark of all ADs is the black or dark gray oil streaks from the exhaust stacks on the plane's sides. At the end of a flight I climb out of the cockpit as usual, but before I can safely make my way off the right wing, my feet slip out from under me and I slide/fall to the pavement – scraping my shin – yep, that's it. Please, no laughing!

Our "extra" instructor left us near the end of our training to start training a new flight of six students. His new students make it through the first two hops, one landing each hop. The next familiarization flight is to an outlying field to practice several landings (8 or 9). Two of his students have fatal crashes while he is on the ground trying to talk them through the landings. When he brings the remaining four students back to Cabinass, he again lands first to be able talk them down – there is another fatal landing accident.

The rumor is that these students have never flown the SNJ during basic training. What the truth is, I do not know. I always think of the

AD as a big powerful SNJ – empty, it weighs about 3 times as much and has about 5 times as much horsepower as the SNJ. I had no problems learning to fly it. But, you certainly HAD to stay ahead of it – especially during takeoffs and landings!

Our flight section, which has last flown together on the 28th, and other flight sections from Cabinass together with flight sections from other local area training fields become naval aviators and receive our Wings-of-Gold during a ceremony at the Corpus Christy Naval Air Station, on Friday, 30 November 1956.

Jackie pins Navy Wings-of-Gold on my uniform during this ceremony.

I have requested an Attack Squadron on the West Coast. My orders came about the 29th to report to the Commanding Officer of Carrier Airborne Early Warning Squadron Eleven (VAW-11), NAS North Island, Coronado, California. I know "VA" means an attack squadron, and I am told the "W" means it is an "all weather" squadron – which makes me happy! But, no one knows what "Airborne Early Warning" means for an all weather attack squadron.

6
VAW-11 – Pre-Deployment

AD-5W of ATG4 Detachment KILO of VAW-11 squadron in Atsugi, Japan (February 1958) (By author.)

The first time I see an AD-5W, the version of the Skyraider I will be flying in my first squadron, Carrier Airborne Early Warning Squadron Eleven (VAW-11), is on Tuesday, 1 January 1957, before I report aboard the next morning. I nearly cry! The big covering of the radar antenna below the center of the wings makes it look like a pregnant guppy fish. Hence, Guppy is a nickname for this plane.

I wanted to go to an attack squadron, and be in a single seat version of the Skyraider. Here is this plane that looks like it is pregnant, and carries crewmen. I now know what "Airborne Early Warning" means for this *non-attack* squadron. I swallow hard and remind myself that it will only be for two years.

I quickly find myself back in ground school on the base to learn all the systems of the AD-5; which are not the same as the AD-1. A sister squadron, VA(AW)-35, on base flies the AD-5N, the plane carries crewmen; but at least their mission is all weather (AW) attack. Pilots new to the AD-5, from both squadrons, are in this class.

The first time I fly in the AD-5W is Monday, 14 January 1957; I am in the right seat where the Air Controller (AC) would normally be. The purpose of the flight is to show me the local flying area, explain the procedures of the squadron and NAS (Naval Air Station) North Island, and the Federal Aviation Administration (FAA) regulations in the San Diego area pertaining to arriving and departing from North Island.

First Night Flight

On my first night flight from North Island, which is after five daytime familiarization flights, Darrel Westbrook and I are flying as a two-plane section. Darrel is another new pilot who reported aboard about the time I did. We decide to fly a little round robin to Los Angeles then Palm Springs and back to North Island before we practice our required night touch-and-go landings on this familiarization flight.

The weather is clear when we takeoff. As we get to LA I am amazed by the amount of light that is emanating from below. For additional practice, Darrel and I are flying formation, separating and rejoining with a different leader. Over LA we are flying separately, as we turn towards the east I lose sight of his plane. So I hang back, waiting till we get further out of the light before trying to join on him. I call him and ask where he is, relative to a landmark. He is still in front of me. As the sky starts darkening again, I see wing and tail lights in front of me and start to work into position to join on the plane. The lights seem too far apart, but I move in closer to get a better look. It is an airliner!

Obviously, I brake off. Soon I see his plane further to the right as the sky continues to darken. I join on him. Shortly after we turn for home over Palm Springs we get a call from the tower at North Island saying the fog is coming in off the ocean and that the field will be *closed* in about 15 minutes!

We both go to normal rated power (maximum continuous power – but less than takeoff power) to get back to North Island before the fog closes the field. We just make it back in time! Our flying time is only one-hour.

About six months later, Darrel tries to join on a star! This has been tried by a number of pilots – and I can see how it can happen on a dark night with both the running and formation lights set on dim. Fortunately, I never tried to rendezvous on another airliner or a star.

Mat Landing

North Island has two long runways at a right angle to one another for the jets and an asphalt "square" mat between the runways for the ADs at the station. The two runways cross near their respective ends, so the asphalt mat is fairly large. The runways are always used during inclement weather and at night by all aircraft. However, new AD-pilots are to use the runways until comfortable with this version of the AD. After a few familiarization flights I have transitioned from using the runways to using the mat for takeoffs and landings.

I am also flying crewmembers on their training flights – an air controller (front right seat) and an aviation electronics technician (directly behind me – but separated by a bulkhead, I cannot see him). On this flight the air controller is learning to use the APS-20 radar and commands for aircraft intercept missions, etc. Two ADs of VAW-11 fly, one as the "target" and the other as the "interceptor", then the planes swap roles – when both planes have full crews.

When landing towards the west, one has to pass over a two-story wood frame building on the air station to reach the asphalt mat. There is a taxiway between this building and the mat. In fact, to takeoff, one merely turns off the taxiway onto the asphalt mat and takes off, after being cleared by the tower for takeoff. One side of the mat is used for landing and the opposite side is used for takeoffs. Remember landings and takeoffs are always in the same direction – and in the direction being used on the main runways. This asphalt mat is used only during daylight and in good weather.

On this flight with a full crew, I have a Chief Petty Officer in the rear as the radar technician. From this rear position he has a small window that looks out the left side of the plane. Landings at this time are toward the west; so after passing over the two-story building I dive a little for the mat, pull the nose up for a good three-point landing (the only way an AD is landed in the Navy).

As I recall it is a warm day and I encounter a more than normal ground effect – i.e. the plane is floating – flying level - about 12 to 18 inches above the mat.

Protocol dictates that there is never any talking by the crewmembers during landings and takeoffs – unless they are reporting an emergency. On the intercom the Chief says, "That was the smoothest landing I've ever had – that was *great*."

I say nothing – we are still flying – at the full stall, which I know is coming; we drop like a ton of bricks onto the mat in the three-point attitude. The landing is safe, but it is the hardest I ever made (before and since) on a land station!

Non-Flying Duties

Upon reporting aboard at 0800 on 2 January 1957, I learn of my promotion to Lieutenant Junior Grade (LTJG), as of 3 December 1956. I am assigned to be the Assistant Ground Training Officer in the Operations Department, which is headed by a Commander (CDR). The Ground Training Officer is a Lieutenant Commander (LCDR) and my main job is to keep the Training Board up to date. A little pin is stuck in the board under the heading of the training received for pilots, air controllers (AC) and aviation electronic technicians (AET) that are new to the squadron. The students are divided into three columns – pilots, AC and AET. Down the left side of each column are the names of students; grouped according to assigned detachments. In this abnormally large squadron (compared to a normal VA squadron), personnel deploy aboard aircraft carriers in detachments rather than the whole squadron deploying aboard a carrier.

Instrument School

From Monday, 23 April through 8 May 1957 I am back in instrument school at North Island flying the T-28B again. I think the school was a detachment of the Fleet All Weather Training Unit Pacific. All new pilots and certain pilots returning to operational flying in the greater San

Diego area are sent through it. They really throw the book at you in this school, so it is nice to know you already have your wings! Most of my flights are slightly over 2 hours, but my final check ride is 3.7 hours. Again I do quiet well.

Bailout

Yes, I have to bailout of an AD-5W!

One day, all new arrivals that are in an operational flying status for the first time, have to put on their flying equipment and report for bailout practice. An AD-5W at the end of one of our aircraft parking lines is secured to the pavement with tie-down lines and has its tail wheel placed upon a platform so that it is in "level flight." The first person, a pilot, starts the engine and goes through the initial engine checks. When the engine is warmed-up he revs-up the engine with the throttle to the designated limit and then dives out head first for the middle of the wing, and is caught in the net behind the wing – as we had done in basic flight training from a SNJ. The second person, who had climbed aboard from the right wing behind the first pilot reaches over and retards the throttle to idle; then moves from the right seat to the left (pilot's) seat as the third person climbs aboard from the right wing and sits in the vacated seat; and etcetera until all have bailed out.

I think it odd that the student officers, who are finishing their training to become Naval Flight Officers (Air Controllers), also dive out the pilot's side rather than "their" side of the cockpit – the right side of the plane. Likewise, the aviation electronic technicians, who fly in the rear seat behind the pilot, dive out from the pilot's seat. Yes, there is an ambulance standing by in case of an accident.

As you may recall, the right side of a propeller-driven airplane is the "wrong" side to bailout, if you have a choice. Also, the aviation electronic technicians never have a chance to miss getting hit by the stabilizer when they bailout of an AD-5 (AD-5W, AD-5N, AD-5Q, etc.) – just not enough space to allow them to fall clear of the tail. Not sure what the technicians are ever told. I certainly never mention anything to them about this!

About six months later when this exercise is being performed again for new arrivals, a non-pilot, a student officer I think, doesn't follow directions and dives to miss the wing. He hits the stabilizer and is taken away by ambulance. As I recall, he drops out of the NFO training program and leaves the squadron.

Compass Rose

The magnetic heading of an aircraft is the compass reading plus the deviation, which may be either plus or minus on a given heading. The deviation is determined by rotating (swinging) the magnetic compass through 360 degrees at a location where magnetic lines with zero deviation are known. At an airfield these lines are painted on the surface about every 30 degrees and the resulting pattern is called the "compass rose." It is located at a remote site on the field. Since the magnetic compass is attached to the aircraft, it is the aircraft that must be swung on the compass rose. The plane must be carefully aligned on *each* heading of the rose, by taxiing.

In VAW-11 the selected pilot taxies the designated plane to the compass rose and meets VAW-11 ground personnel there. The personnel will provide the pilot taxi directions. When signaled, the pilot carefully reads the magnetic compass and notes the heading on a sheet of paper. From these readings the deviations verses headings of the compass will be printed and posted below the magnetic compass.

This swinging must be accomplished on a periodic basis and whenever maintenance on the plane has changed its residual magnetic field. Thus deviation is the local distortion of the earth's magnetic field by the plane's residual magnetic field, which varies with respect to the heading of the plane. Ships must also be "swung" to calibrate their magnetic compasses.

As a first year junior pilot, I am tapped a few times to provide this service for the squadron. Prior to one of these "missions" I am in the ready room with my flight suite and life vest on – as worn for a normal flight. The SDO (Squadron Duty Officer) calls me to his desk and tells me to report to the line shack as the compass rose pilot.

At the line shack I am informed which airplane to take; all of our planes have a three-digit number painted on both sides of the engine cowl. I walk out to the plane and give it a quick pre-flight (yes, I know I am not flying); but before any pilot moves a plane he wants to look it over to see if all attached parts are in fact attached, there is no loose gear, liquids aren't dripping from the plane, etc.

When finished, I climb into the pilot's seat and fasten my lap and shoulder harness at the single point of connection. Once I set the switches and valves in their proper positions for starting I wait until the head linesman signals they are ready for me to start the engine. His signal means no one is near the arc of the prop and a second person is holding a fire extinguisher for instant use.

Once the engine is started, I place the harness in its locked position and check the instruments for normal functioning. These procedures are done automatically, without really thinking about them after starting the engine. The locked position has the harness tight across the pilot's body, holding him against the back of his seat.

Then I signal to pull the main landing gear wheel chalks. Once these are pulled the head linesman begins giving me taxi signals to clear the other parked aircraft and head for the taxiway.

Just prior to the taxiway I stop the plane and unfold my wings, when signaled. I call the tower for permission to taxi to the compass rose. Upon receiving taxi clearance I start taxiing towards the rose. Since I am in a taildragger I am continually doing S-turns on the taxiways to the remote spot of the compass rose. The purpose of the S-turns is so you don't hit another plane or vehicle – you can't see directly in front of the plane because of the cocked-up attitude of the fuselage in a taildragger. The engine blocks the forward view.

When I am about halfway to the rose I suddenly have tremendous pressure on my chest and I can't breath! I have no idea what is happening to me – I'd never had an attack like this before!

I stop the plane immediately and look down at my chest – *surprise*; my life jacket has inflated on the left side. I quickly jerk the single-point attachment lever, but it is partially jammed because of the increased tension on the shoulder harness caused by the inflated life jacket being restrained by the harness. I apply extra force to the lever, and the lap belts and harness finally go flying off.

Hurray, I can breathe again!

How the left toggle got hung on something that resulted in its being pulled I do not know, though I had heard of this happening with other pilots and crewmen. Fortunately, I never have it happen again.

Rescue Flight

Another mission given to me because I am a junior pilot is to "rescue" a more senior pilot who has put his plane down at an emergency-landing site along the coast, well north of San Diego. He was flying as a "target" airplane, without crewmen on an AEW (Airborne Early Warning) training flight, when his engine began having serious problems.

The SDO doesn't know which emergency field the pilot has used. There are several along the coast, mostly old military outlying training fields no longer in use. However, he is *sure* that at whichever emergency field I find an AD-5W that will be the right field!

So off I go to look for his plane.

As I recall, he is at the field that the SDO and I have guessed would be the one. As I fly low over the field I see him standing next to his plane waving. Fortunately he has been able to taxi his plane off the runway onto a taxiway. I look at the grass (weeds) and trees to see which way the wind is blowing and select the proper runway for landing. I taxi to his plane and stop nearby. With my engine idling he climbs into the right seat. We shake hands and he thanks me for picking him up.

I taxi back to the end of the runway I had landed on, turn into the wind and finish my takeoff checklist. Since the runway is short with a mountain ridge at right angles to the runway fairly close to the opposite end, I put my flaps full down.

At most land stations takeoffs are made with the flaps up, on a deck launch from a carrier they are *always* full down. Full flaps provide a shorter takeoff run, because the minimum flying speed is reduced, and a higher *initial* rate of climb, than a flaps-up takeoff.

Since he is a more senior pilot I say, "Guess I ought to use full flaps," to get his recommendation, if he has one. He says nothing, which I take to mean that he agrees with my decision.

Shortly after lift off I turn left so we will not have to clear the ridge in front of us – which I know the plane cannot do. However, this puts us climbing in a canyon with mountain ridges above our altitude on both sides of us. I realize the "initial" climbing phase is over and it is time to increase our rate of climb; so I begin "milking-up" the flaps.

"Milking-up" is incrementally raising the flaps. Each time the flaps are raised a little; the air speed increases a little, which in turn increases the rate of climb a little. During a maximum rate of climb, if one just raised the flaps in one step, the plane would actually lose altitude; and if not allowed to lose altitude it would stall (resulting in more altitude loss, if not a crash).

We clear the mountains with room to spare before the canyon ends, which would not have happened if I had left the flaps full down. It is good applied training for me.

I am glad that I have spent some time looking over the performance tables and curves in the back of our aircraft flight handbook. I always feel not enough pilots study these back pages. I also feel that my aeronautical engineering education helps me to more easily understand these pages.

What surprises me most is that my passenger never makes any comments during the climb out. I assume I must have handled it the same way he would have, which makes me feel good. This is my first experience of low flying near mountain ridges.

Our squadron sends a vehicle with maintenance personnel to fix the plane's engine and a pilot to fly it back.

Instrument Failure

One morning, the ceiling is too low for using the mat, so all the ADs have to use the runways. The duty runway is 27 (heading 270 degrees – West), which ends just short of the water. This water is the narrow neck of water that connects the port of San Diego to the ocean. On the other side of this strip of water is the Point Loma Peninsula, which runs north and south. Therefore, when taking off on this runway one executes a tight turn to the left (south) in order to not fly into the peninsula! It is too high to fly over on takeoff – at least in our aircraft.

This particular day, the peninsula is shrouded in clouds. As soon as I become airborne I initiate a left hand climbing turn, shallow at first, but steepening as altitude is gained. During the turn I enter the clouds and begin flying on instruments. My gyro horizon, gyrocompass and turn-and-bank indicator all indicate I am turning left.

However, upon rolling out of my turn to fly 180 degrees, my gyrocompass keeps showing a slow turn to the left. Not knowing for sure what is wrong, I keep climbing and initiate a shallow turn to the left based upon my gyro horizon and turn-and bank-indicator – to make sure I do not fly into the peninsula. I cannot trust my gyrocompass for a good heading! Soon I successfully come out on top, in a shallow left climbing turn. This is my first experience of instrument malfunction or failure.

Hunter Killer Flights

In theory, when a VAW-11 aircraft is flying an ASW (Antisubmarine Warfare) mission, we, the hunter, have a killer aircraft – like an AD-5N from VA(AW)-35 carrying depth charges - flying formation on us. When the air controller spots a radar image of a periscope, he quickly determines the course and speed of the submarine. Then he directs the pilot, as required, to turn to a non-closing, non-threatening direction.

Once settled on a new course, if required, the air controller informs the killer aircraft of the range and bearing of the submarine. The killer plane then descends directly below the hunter, maintaining the same heading and horizontal speed of movement.

At this time, it is assumed that the radar on a periscope will not have height discrimination capability, and thus will see only *one* radar image that is flying in a non-threatening manner.

Once the killer plane is just above the water he asks for the attack heading and distance to the submarine. At this distance the killer plane should be below the sub's radar horizon and not seen turning for the attack. In theory, by the time the very low flying plane is detected by the sub's radar it will be too late for the sub to escape.

In the early stages of hunter-killer flights this is the way it was accomplished.

Once in awhile we have one of our submarines be an "enemy" submarine so our controllers can see the radar image of a real periscope. Of course the practice area is carefully defined and a common radio frequency with call signs for contact has been established.

Sometimes we (VAW-11) perform this exercise with one aircraft and other times we have two aircraft – playing the hunter-killer game. With one aircraft, the hunter becomes the killer, and the controller gives heading information to the pilot to fly directly over the periscope. The pilot is supposed to see the periscope pass under the aircraft for verification of the "attack."

At the conclusion of one these exercises, I have forgotten whether one or two planes are involved, the submarine surfaces and I can see people on the top of the conning tower waving at us. So, being a good sport, I dive on them with my wings "waving" to them as we fly very low by the coning tower. I pull up rather sharply, with nice vapor vortices trailing from the wing tips!

However, I pull a little too hard and the plane stalls!

This is known as a "high-speed" stall because most stalls occur at low speed. A stall means the angle of attack is too great for the given airspeed and all the lift of the wing is lost. It may be one wing (in terms of left and right) or both. It is just my right wing that stalls, so I quickly recover with minimal loss of altitude and turn it into a right banking turn.

My recovery is so quick and smooth that neither my controller nor electronics technician in the back ever realize I have accidentally stalled the plane! If my memory is correct, this is my first and last accidental high-speed stall.

Needless to say I would have *really* been embarrassed had I splashed my plane along side a submarine – though the rescue would not have taken long; assuming they didn't laugh too long.

FCLPs

I have been assigned to Detachment KILO, with a scheduled deployment aboard USS *Hornet* (CVA-12) in early January 1958. This means we must be carrier qualified in type, AD-5W, well in advance of our departure date.

This *Hornet* was commissioned on 29 November 1943 as an Essex class carrier (CV-12) with a straight wooden flight deck. Since then she has undergone several modernizations, including the installation of an angled deck (12 degrees off-center) for safer and easier landings, primarily for jet aircraft, a British innovation.

In addition, the Mirror Landing System (MLS) was installed, another British development. This provides a visual indication to the pilot as to his altitude relative to the desired approach path. Since the MLS is relatively new, AD pilots, at least those flying in the Pacific Area, are required to qualify on *both* paddles and mirror. Thus, if the mirror system fails, there is a backup method for recovering the Skyraiders.

Prior to carrier qualifications, all pilots must complete Field Carrier Landing Practice (FCLP). This is accomplished by simulating carrier approaches at a remote runway with a Landing Signal Officer (LSO). One must demonstrate numerous satisfactory daytime field carrier landings to the LSO before being permitted to fly night FCLPs. However, the daytime FCLPs are still flown from time to time in conjunction with night time practices. The LSO determines if a pilot is ready for carrier qualifications.

The field carrier landing is really a touch-and-go landing (with the hook up); but the approach is flown differently - lower and slower than for a normal runway landing. An AD paddle-controlled landing circuit is flown at 200 feet while a mirror-controlled circuit is flown at 300 feet (above runway height).

AD-5N receiving a "cut" at Brown Field in 1954 (US Navy Official)

Another difference between a carrier landing and a runway landing, which cannot be practiced on a runway, is the control setting by the pilot of the tail wheel. When a tail wheel is locked it cannot swivel. When landing on a runway the tail wheel must be set to "locked", but it *must* be set to "unlocked" for a carrier landing.

Both FCLPs and carrier landings are accomplished under the supervision of a LSO, who stands in his area near the approach end of the landing area, facing the landing aircraft. During a paddles approach, the LSO uses paddles (short handled "tennis rackets" – often with colored cloth strips in place of horizontal strings) to signal desired corrections and either a "cut" or a "wave-off" to the pilot.

LANDING SIGNAL OFFICER (LSO) SIGNALS

US Navy LSO Manual

The LSO stands with his feet apart and arms spread horizontally, when the approach needs no corrections ("roger pass"). In conjunction with the paddles, brightly colored cloth strips usually go up the front of the legs and torso to the neck, and from the left wrist, across the chest and to the right wrist of his flight suit, to improve visibility during the day.

At night these strips are replaced with lights, both on the LSO's suit and paddles. On final approach the LSO can signal up to 14 desired corrections with his arms and legs, plus 2 *mandatory* commands with his arms: "cut" and "wave-off," as shown on the previous page.

There are six basic components to a FCLP or carrier approach and landing (1) Flight Pattern, (2) Altitude Control, (3) Air Speed Control, (4) Attitude Control, (5) Final Lineup, and (6) Landing Technique. The LSO tries with the above signals to communicate to the pilot how to improve on any or all of the first five in order to earn a "cut" for a safe landing.

With the mirror, all communications (except the "cut" and "wave-off") to the pilot by the LSO must come over the radio. The pilot never talks to the LSO during an approach, except to make specific reports. At any time on final that the LSO thinks the landing cannot be safely made, he signals a "wave-off."

On one day FCLP session at Brown Field, an auxiliary field near North Island, the running of my engine seems to be getting a little rougher on each pass. On this one pass I get a "cut", make a landing, add power to takeoff and when airborne my engine almost quits, so I simultaneously notify the tower of my intensions and pull all the power off to land again on the far end of the runway. This is the first and last time I ever had the thrill of landing twice on the same runway during the same pass!

Night time adds another big pucker factor to flying FCLPs!

From the 180-position (flying down wind and opposite the end of the runway) onward you are on the approach to a landing. Brown field is in the country, away from all small towns, so at night there is nothing to see, except the runway lights (which are set on dim) to give yourself any reference to the ground. Thus during the up-wind, down-wind and the first part of the approach you are flying on instruments at a very low altitude.

After flying through the 90-position you have the runway lights to give you some guidance, but you must also be on instruments. Around the 45-position you must follow the information being provided by the paddles and your instruments. With the MLS you don't pick up the ball until about the 30-position – you must be within the cone of the reflection of the "ball" in the mirror.

During the day you can glance at your instruments from time to time, but during the night you gradually transition from looking at your instruments all the time to at the very end, looking at only the total scene - LSO/MLS and landing area on the runway/deck. You cannot stare at the landing area, and make a successful landing!

On our way to Brown field to practice FCLPs we always fly in a tight four-plane formation, but on the way back to base we fly in a very loose formation and pass very close to our home. When returning from night FCLPs I always blip my engine to let Jackie know I am heading back to North Island, and will be home presently.

Big Decision

I receive a letter from the Chief of Naval Personnel informing me that the end of my obligated active service time of 3 June 1958 (this has been increased a year because of my flight training) is less than a year away and I must submit a letter requesting retention as a permanent officer in the regular navy by the end of December, if I want to be considered for retention by the next annual selection board. After a couple of months of considering my options, including visiting an aircraft manufacturing company in San Diego and talking with an aeronautical engineer in this company, I submit my letter for retention as a permanent officer in the regular navy, on 26 December 1957.

7

VAW-11 Cross-Country Flights

Background

Before a pilot flies a military aircraft he must sign for it, in much the same way as one signs for a rental car. However, before signing, one reviews the "yellow sheets", which normally go back about 10 flights or so, for the assigned aircraft. During this review one sees exactly who has previously signed for the plane, what the complaints have been by the previous pilots and what the remedies have been and who signed off on them. In addition, any other work that has been performed should be noted, and signed off by the person responsible for the work. Also, if critical-to-flight work has been performed on the plane one must see that a test flight has been performed and by whom; unless you are the test pilot.

If the pilot "gripes" and maintenance fixes indicate that the plane will not perform satisfactorily on the assigned mission, one may ask for another plane. If no other plane is available, one may ask the Squadron Duty Officer or Operations Officer if it is worthwhile to try and fly the mission. The pilot may "Down" a plane after the preflight or after conducting checks and tests from the cockpit prior to flight. (By the way, from a book I just read that was written by a current/recent naval aviator, these plane records might not be called "yellow sheets" in today's Navy)

The person who signs for the aircraft is responsible for the plane, both legally and financially! Therefore, if a LTJG signs for an airplane but has a CDR pilot flying with him, it is the LTJG who is in charge of the airplane and not the CDR. The only exception to this is when the other pilot is the Commanding Officer of the squadron. He, and he alone, is responsible for all the planes in the squadron and all the personnel.

There are a lot of flight requirements Navy pilots have to accomplish during a year, such as: minimum nighttime hours, minimum instrument hours (simulated/actual), minimum number of instrument approaches including a minimum number of GCA (Ground Control Approach) instrument approaches, minimum number of total first-time hours (monthly and annually), and an instrument check ride (pass/fail). For carrier pilots, these also included a minimum number of arrested landings (both day and night) and catapult shots (both day and night) every 6 months, to remain carrier qualified. In addition to these requirements, a pilot has to take at least one cross-country flight a year under Instrument Flight Rules (IFR).

In the Navy, when flying in daylight at or above 10,000 feet in an un-pressurized plane such as the AD, everyone must use oxygen by wearing an oxygen mask. Likewise when flying at night in these airplanes one must use oxygen at and above 5,000 feet to improve night vision. Prior to these flights the oxygen system of the plane is always tested by each crewmember with his mask on. Crewmembers normally take their masks off prior to takeoff when oxygen is not going to be required for awhile.

Round Robin to Austin

In the VAW-11 ready room LT Zimmerman approaches me about going with him on a cross-country flight to Austin, Texas. Knowing I need the cross-country flight and that I can visit our parents in Temple, I say okay. We are to leave the coming Friday, 7 June 1957, morning and return Sunday evening.

As is normal procedure, since Zimmerman has signed for the aircraft he will fly the first leg of the flight to El Paso, Texas. He files an IFR plan for the "obligatory" annual cross-country flight. As I recall we have no actual instrument flying enroute to El Paso. We land at the El Paso municipal field and taxi over to the aircraft service area.

As our plane is having the fuel, oil and the oxygen tanks refilled, we have a quick lunch before we go to the weather office to obtain our forecast weather from El Paso to Austin, and have it recorded on our flight plan. The forecaster tells us we better hurry, because a dust storm is approaching the field from the west. Since we have come from the west and have not noticed a dust storm, how bad can it be?

We walk out to our plane, and looking off towards the west, up-wind, we see in the distance what appears to be a very dark, deep "fog-bank." The linesman tells us we need to hurry.

I am flying the plane, so after a quick preflight of the plane I climb into the pilot's seat. (The preflight is never very long on a plane you have just been flying.) I start the engine and before I can call the tower for our flight clearance to Austin, the linesman begins frantically motioning for me to taxi. So I give him the signal to pull the wheel chocks.

He has me turn right and taxi only a few feet *downwind,* before he signals me to "Stop."

Surprisingly, he gives me the "cut-engine" signal! (Stop my engine.)

On our radio, which is tuned to the ground control frequency, we are listening to a FAA air controller telling the pilot of a military jet training plane, which is holding near the approach end of the runway, the following: "Your clearance to "XXX" is as requested and you will receive this clearance in the air if you *immediately* takeoff, otherwise you should forget about the flight." (Legally, a pilot *must* receive an IFR flight clearance and acknowledge it *before* takeoff – I had never heard of a FAA air controller willing to bend this rule – though under this circumstance I think it is a great decision, since "XXX" is eastward.)

Since I now understand how rapidly our weather is deteriorating, I immediately cut the engine as signaled! It is obvious the linesman realizes we are not going to be able to takeoff; but he wants our plane positioned facing downwind to protect the engine as much as possible.

I watch the jet immediately pull onto the runway and takeoff. As I hold my breath, I see the plane just barely clear the wall of dust, which is several hundred feet high, not far beyond the edge of the field. I think it wouldn't take much sand in a jet engine to severely damage it.

As I look back towards the linesman he is standing beside the cowl of our engine holding a tarp he has retrieved. Doing his best to protect our plane he throws the tarp over the engine. It will cover our carburetor air-intake, which is a few feet in front of our windscreen. He gets a couple of ends tied before he has to run for cover.

We close our canopies and sit there as the wall of dust rapidly approaches and completely envelops us. Neither of us has ever seen any dust storm like this, but obviously the linesman has!

It seems a long time before we can see anything around us. We remain in the cockpit until the visibility almost returns to normal and we see the linesman approaching our plane. We open our canopies, climb out, and thank him for his quick action to help protect our plane! We assist him in its cleanup. As I recall, the linesman uses a high-pressure air hose to completely remove all the sand and debris from our plane.

Several hours goes by before our plane is again ready for flight; but the field is still closed while it is being restored to an operational condition. We have to wait so long that we become hungry, so we have another meal and then wait some more. Zimmerman needs to be in Austin the next morning, so we do not consider spending the night. When we are informed that the field will reopen soon, we return to the weather office to acquire our forecast update.

The night takeoff and flight to the Air Force base near Austin presents no more problems. It is about 0200 in the morning as I commence a standard military approach, as cleared by the tower, by flying down the duty runway at about 250 feet. I make the break just before the end of the runway and complete the landing checklist as I near the 180-position. I am preparing to call, "Navy 33761, wheels down, flaps down," but a final check of the wheel position indicators shows a "barber-poll" in the window for my right main gear, while the other two wheel indicators were "in-the-green" – as all three should have been!

So, I call, "Navy 33761, is waving off with right main gear indicating *not down.*" I raise my flaps and climb to about 1,500 feet as I come back around to the runway heading, and then entered a dive to 200 feet followed by a sharp pull-up to exert enough positive Gs to complete the extension and mechanical locking of the right main gear – if this is the problem.

The "barber-pole" indicator remains, so I ask, "Navy 33761 requests a low-fly-by for the tower operator to see if our gear appears fully down?" He agrees, so I lower flaps to full down in order to fly as slowly as possible as I maneuver to fly our right wingtip *very* close to the tower, at eye level to the observers. Since we can see the whites of their eyes, we know they have a good look at our right main landing gear.

The tower operator reports: the gear looks fully down.

Since Zimmerman has signed for the plane, I ask him what he wants to do. (Implicitly I am asking him, do you want to land with the gear down or up?) If we *know* the gear is not locked down, we have to land wheels-up. There is less damage to the plane and it is safer for the crew to make a wheels-up landing than to have a main landing gear collapse upon landing, sending the plane cart wheeling off the runway! This lesson was well learned during WW II.

On the AD, the last actions of the main landing gear upon extension are the mechanical locking of the struts, which may be accomplished with hydraulic pressure or positive Gs, *and* the plunger of a micro-switch being pushed inward by the strut to close the circuit that indicates on the instrument panel that the gear is *locked* down. Since I had hopefully pulled enough positive Gs to have completed the gear extension and the tower personnel think the gear is fully extended, the odds are we have a micro-switch failure and *not* a strut-locking failure.

He says, "Let's give it a try."

So I report to the tower operator that we will be making a landing with an "unsafe" gear indication. Before we are back to the 180-position for another landing approach, several fire/crash trucks are moving towards the duty runway with their flashing red lights and sirens blaring; which is standard procedure for this situation.

I am wondering how many people we are waking-up this morning!

I make a two-point landing - left main and tail wheel – prior to putting weight on the right main gear. This is the way it is done by a Navy taildragger. The right main touches down and we continue straight down the runway. We stop on the runway to give the crash crew an opportunity to check that the right gear is in fact *locked* down.

Planes have been known to have an unlocked main gear collapse upon trying to turn off the runway. Thus, the crash crew is supposed to go under the plane's wing and check that the locking mechanism is in fact in the locked position.

With our engine idling they approach the gear from the right wingtip. They soon emerge at the wingtip with a thumbs-up. I admire their bravery to go so close to a turning propeller, especially at night, to make this check for us.

I turn off the runway onto the taxiway that leads to our parking place, in accordance with tower instructions.

This is my first, but not my last, landing with an "unsafe" gear indicator.

My parents drive down later in the day to see me, but because of our late arrival, I do not return with them to Temple.

The following morning Zimmerman is to fly our plane to El Paso. The Air Force maintenance personnel have replaced the faulty electric micro-switch and topped off our fuel, oil and oxygen.

We arrive as scheduled in El Paso with no problems enroute. Again we eat and receive our forecast weather for our flight while our plane is being serviced. I am flying the last leg to NAS North Island.

Before starting the engine, one of the electrical switches that you put to "ON," is for the "fuel boost pump." The standard fuel pump is engine driven, so before the engine is started there is no fuel pressure from this pump. During the takeoff checklist, this check is performed in the run-up area near the end of the duty runway; one switches the fuel boost pump to "OFF" to check that the engine driven fuel pump is working properly.

In the run up area, when I switch it "OFF", the engine starts to die. I quickly switch it back to "ON" and the engine runs normally. After I have finished the takeoff checklist, I again switch the fuel boost pump to "OFF." The engine again starts dieing. I quickly switch the boost pump back to "ON" and the engine runs fine.

I inform Zimmerman that the engine will not run with the fuel boost pump turned off.

The pilot's flight manual says that you **will** *not* takeoff if the engine will not run with the fuel boost pump "OFF." As a *required* precaution, all pilots takeoff and land with the fuel boost pump switched to "ON" – following the takeoff and landing check lists. These two pumps are connected in serial, not parallel mode, so the boost pump output feeds the engine driven pump.

Zimmerman says that because the fuel is very cold and the day is very hot, the resulting excessive vaporization is more than the fuel pump can handle without the help of the boost pump. That it would be okay after we are airborne. Remember, he has signed for the plane; and he sounds very knowledgeable.

So, I trust his judgment. I have heard of vaporization before, such as vapor lock, but have never experienced it. After we get our flight clearance for North Island followed by our takeoff clearance, I take the duty runway and takeoff.

As normally done, as we pass through about 4,000 feet I switch the fuel boost pump to "OFF" – the engine immediately starts to quit; so back to "ON!"

After we level off at our assigned altitude of 14,000 feet, I again test the engine without the fuel boost pump, with the same results. Zimmerman says the fuel just needs more time for the fuel temperature to stabilize.

I start wondering if he knows what he is talking about and how long a fuel boost pump will run. Normally the boost pump is only used from starting to shortly after takeoff and from the landing pattern till parked. And of course it is to be used in emergency situations – like our present predicament.

About 20 minutes later I test it again. Hurray, the engine runs fine without the boost pump!

I am not sure what has really occurred, but the vaporization theory sounds correct. Of course it would have been a real coincidence had our engine driven pump just happened to fail during or shortly after our landing at El Paso. However, I *would not* have taken off if we had not just flown the plane!

A few months later when the Wright Engine Company representative is visiting VAW-11 to discuss the care and treatment of the AD engine and to see what problems we are having with our engines, I ask him about the problem we had had. He says he had never heard of excessive fuel vaporization preventing the engine driven pump from performing its function; and he would not recommend taking off under these conditions.

Hope you have not noticed that we had problems only when I was doing the flying!

Round Robin to Waco

Since I did not get to see our parents, I request to fly another cross-country to Texas, this time to Waco, I am doing all the flying. My passenger is Bill Chapman, a student Air Controller who has recently joined our squadron following his initial Air Controller training. He will be designated an Air Controller and receive his Naval Flight Officer (NFO) wings after completing our squadron's training for air controllers. He is assigned to Detachment Kilo, and we do a lot more flying together.

Our flight, on Friday, 28 June 1957, is again via El Paso.

At North Island the weather forecaster informs me that we will have a lot of cumulus activity over the continental divide, but he is not forecasting thunderstorms. In the Navy one cannot file a flight plan that takes you through existing or forecast thunderstorms. I file an IFR flight plan, as required, because of the forecast weather and squadron requirements.

During an IFR flight plan below 18,000 feet the pilot has to fly on the published airways and maintain constant radio contact with the FAA air controllers. At each reporting point on the airway one has to report when arriving, give a time estimate for the next reporting point and the name of the following reporting point. Such as: "Navy 35195 over Able 1147, estimate Baker 1223, Charlie." The response is, "Roger Navy

35195, Able 29.98." The "29.98" is the atmospheric pressure at station Able, which one then adjusts the value in the pressure window of the altimeter; plus or minus the altimeter error, which the pilot has determined prior to takeoff. This keeps all aircraft flying under IFR requirements at the correct height within the local area.

When flying at flight levels, which are above 18,000 feet, the altimeter is set at 29.92 plus/minus the error – the pressure of the standard atmosphere.

The "35195" are the last five digits of the serial number of the aircraft, which has nothing to do with the three-digit number painted on the engine cowl. However, these five serial-numbers are painted in smaller size on both sides of the fin/rudder so a tower operator can see them, and the full serial number is painted in very small size on the sides of the fuselage near the tail, usually below the stabilizer. The full serial number is also in the cockpit for the pilot to see.

One has to arrive over the next reporting point within 3 minutes of the estimate. Too early or too late is a flight violation; however, enroute one can revise the estimate. On these cross-countries we normally cruise at 180 knots true airspeed, which is about 207 mph and about 2/3rds of our maximum speed. This speed gives us good fuel economy and facilitates mental navigation estimates. (Remember the AD was designed to be a dive-bomber, not a fighter. The AD can carry a heavier bomb load than the four-engine B17 bomber of WW II. In fact, the AD can carry its empty weight in ordinance!)

As we approach the continental divide I am flying on instruments. The closer we get to the divide the rougher the air gets. We hear airline pilots talking about diverting our general area. Since we see no lightning, we are not in a thunderstorm - by definition. However, it is raining very hard.

When my 3-minute window for our next reporting point is about to expire, I call FAA and extended it by 3 minutes. The air controller laughs, and gives me his estimate, which is about 5 minutes; based upon the time it has been taking previous aircraft. The tailwinds are much less than forecast. His estimate is correct.

Soon our "rate-of-climb" indicator begins going up and down like a yo-yo while I hold the controls steady. We are dropping 4,000 feet a minute and then rising 3,000 feet a minute – it is really jolting! I think my calm attitude is helping to relieve Bill's fears.

I keep a careful eye on the altimeter, to make sure during the downdrafts that we are still well above the mountaintops. We certainly are going outside the altitude limits of our flight clearance, but I know no plane is near us; so I don't worry about hitting another aircraft. I only have two concerns - staying above ground and the wings staying attached!

Rather than flying steadily I begin trying to help soften the sudden updrafts and downdrafts. On the intercom I am keeping the comments light, inferring I have experienced all of this before and there is nothing to be concerned about. In reality this is my first experience of such a terrific battering! It also turns out to be the worst I ever experienced!

I am *very* glad that we are in an airframe built for dive-bombing. Coming out of those training command dives in our AD-1s we normally pulled about 5 positive Gs. We were not supposed to pull over 6 positive Gs without reporting it. In addition to our real-time G-meter we had a maximum/minimum G-meter that retained the maximum and minimum we reached during the flight, which we set to zero/zero before we taxied to the runway. I had pulled over 6 positive Gs on only one dive during training, which I did report. We did not wear any type of G-suites during our training.

In the AD-5Ws we have no G-meter, but we are definitely pulling less than we had in our dive-bombing training.

Just about when I think this "bronco-ride" will never end, it stops. I estimate it was a good 20 minutes in duration. Finally we are back to typical cumulous bouncing. What a relief!

We are almost to El Paso before we are finally flying in clear skies.

Our trip to Waco and landing at the Air Force base are uneventful.

Dad drove up from Temple to pick us up. My parents are still living on the VA Hospital grounds. The next day I visit Chap and Velma (Jackie's parents) for a while.

The following morning Dad and Mother drive us up to the Air Force base in Waco. They remain until we takeoff. They are standing on a balcony just above our plane as we climb aboard and taxi to the duty runway.

Later Mother says she had never seen a plane climb so steeply after takeoff! To us it was a normal AD flaps-up takeoff.

The trip to El Paso is uneventful and on the ground everything goes well.

We have been cleared to climb to 14,000 feet and then proceed on our way at this altitude to the San Diego area. *Suddenly,* just as we are a couple hundred feet short of reaching our assigned altitude, the engine starts screaming (over-speeding) because the propeller has gone into flat-pitch, producing little or no thrust. Instantly, I pull the throttle to idle – preventing the engine from coming apart! The cause of this predicament is the failure of the propeller RPM-governor.

We are gliding – something I thought I might be interested in, learning to be a glider pilot, but *not now*!

While pulling the prop control (RPM-lever) back from the normal climb position, I hope the RPM of the engine will further decrease. Finally, as the prop control approaches its lowest setting (maximum pitch), the RPM starts decreasing – much to my relief! This means the governor is at least partially working, with an extremely low power setting. Of course, we are still losing altitude.

For good measure, I pull the prop control to its lowest RPM setting, before *slowly* pushing it forward into the normal cruise range. The RPM cannot increase to this setting because of the retarded throttle position.

Now the **big** test! *Gently*, I push the throttle forward to increase the power and RPM. The question is: "Will the RPM increase only until it matches the setting of the prop control?"

As I *ease* the throttle forward the RPM finally stabilizes, and does not increase. Great! The governor is holding the RPM constant and we finally stop losing altitude. We are now near 11,000 feet.

Gradually, I increase the throttle to see if we can reach cruising speed without the governor failing. Success! We are cruising at 180 knots.

For Bill's benefit, I try to act as though this is a common experience.

Not wishing to push our luck any further, I ease the plane into a slow climb without increasing the throttle to normal climb power. I inform the FAA controller we have had a runaway propeller while climbing to 14,000 feet, that we are presently climbing again to 12,000 feet and request a revised altitude of 12,000 feet for our flight plan. Fortunately, he quickly clears us to fly as requested.

Now, if only the governor will function properly until we arrive at our home field.

It does! As a precaution I do not increase the prop setting, as is *always* done under normal conditions, prior to landing.

I explain the problem with the propeller governor and "down" the plane as I fill out the Yellow Sheet. This governor is replaced and sent for overhaul.

This was the first, and thankfully the last, time I ever had a runaway propeller. My reaction to this dangerous situation came naturally - I *had* to give the governor a second chance.

I have never met another pilot who has had this experience while flying a *single* engine plane. However, all of these pilots who have heard of a single engine plane having a runaway propeller have also heard of only two outcomes: bailout or crash landing!

Multiengine pilots have told me that on the rare occasions when they have had a runaway propeller, they have *always* chopped the power and secured the engine; meaning it was not worth the effort to try and "save" an engine in this condition – considering the LOW probability of success.

Oh the luxury of having more than one engine!

* * * *

Please see the Appendix for more information on how the engine of a propeller-driven aircraft is controlled.

8

SURVIVAL SCHOOL & ESCAPE AND EVASION

Prior to deployment to the Western Pacific all aircrew must have the Survival School and Escape and Evasion training blocks checked-off in their records. In the San Diego area the Navy has combined these two requirements into a week of specialized training. In November 1957 I am in my final phase of being prepared by VAW-11 for deployment in January. Thus it doesn't come as a surprise when told I will be receiving this training next week. I am to report at 0800 on Monday at the pick-up spot for NAS North Island participants, wearing and having only what I normally have while flying VAW-11 missions; except I am not to bring my 38 caliber revolver.

A month or so earlier one of our pilots who is to deploy before me returned from this training with a face that had obviously been beaten, including a black eye. All he will say is that he had gotten into a fight with a Marine during the training in the mountains. Apparently he had been caught during some exercise and either tried to escape or refused to obey an ordered. The other pilots and air controllers that have received this training seem to understand and say, "The Marines do not play games with you; they fully carry out their part as the 'enemy' during the exercises!"

I am the only officer from VAW-11 who is attending at this time. I have not previously met the two VAW-11 enlisted personnel who are also attending. We are dressed in our standard *summer* weight flight suits, which are not worn over a uniform, and are wearing our Navy leather flight jackets.

A special Navy bus, with other passengers onboard, picks us up and takes us to the training camp in the mountains east of San Diego. A previous bus full of trainees from various squadrons within the general area has already arrived.

We are driven into a compound that is apparently a Prisoner of War (POW)-type camp. In the main building we are informed in general terms the reason for this training and what we will be doing. The first two days will be on survival and evasion techniques in the mountains, the third day will include an exercise on evasion and escape. The Marine commander of this unit says, "Those who get captured during this exercise will learn more than they desire of how badly POWs are treated and the rest of you will see very realistic interrogations of the 'prisoners' before your departure." The last two days of the week will cover survival and rescue techniques for beaches and the ocean.

The head instructor for the first day's exercise takes the podium and explains the scenario: we have crashed in enemy territory in a remote area and our task is to reach a predetermined helicopter rescue point for this area, prior to twilight. We are shown a topographic map of the local area with the crash site and helicopter rescue points illustrated. We cannot use a compass because we are to learn to land-navigate by way of a topographic map and the terrain we see. Our task is to vacate the local area rather rapidly before enemy soldiers arrive. However, we have to be aware that enemy aircraft might be used to search for us, so in our evasion we have to remain covered most of the time. There is a road in the valley that will be used by the enemy for patrols, so we must remain clear of this area. In addition, we are to stay away from ridgelines where we will be silhouetted against the sky.

Thus, we should strive to travel about 3/4ths the way up the side of the ridges to provide the best evasion options. Remember, the worse the terrain, the safer we are from capture – but we have to balance this with vacating the crash site and arriving in time for a possible pick-up prior to nightfall.

Outside, we are issued a web-belt with a canteen of water and then split into four groups, with two instructors per group. Squadron mates are separated.

In our group we look at the contour map and relate that depiction to the topography we see and tell the instructors which distant landmarks will serve as our references as we hike across the rugged terrain. So, off we go at a brisk pace.

Our instructors occasionally adjust our heading in order to teach us different hiking techniques appropriate for different topography. Periodically we have to locate our position on the map and identify new landmarks to guide us. Each group travels a slightly different route to the common destination, because of the differences of the instructor's preferences or techniques of teaching.

It has snowed in this area the previous night. We are instructed *not* to eat the snow because it is not good for our systems. However, collecting snow for drinking after it has melted is a good idea – but we will not need to do this because they have furnishing fresh water for us.

Upon arriving at the rescue point, each group is split into new smaller groups of about 10 people each. A parachute is given to each group in order to make a tent, and each person is given half a parachute to help keep them from freezing at night. Our instructors make suggestions on how to make a tent from a parachute and warn us we should keep a fire going in the center of the tent throughout the night!

We gather wood from below the trees in the local area. The fire is built on the ground in the center of the tented area. There is a hole in the top of our tent for the smoke to escape. Fortunately, we have some "boy scouts" in our group who keep the fire going through the night.

One side of your body is warm while the other side is freezing, so very little sleep is obtained while we keep adjusting our orientation to the fire during the night. My canteen is by my side next to the edge of the tent. By morning the water in my canteen is frozen! It is by far the most miserable night I've ever had to endure!!!!

Early the next morning our instructors return and we have a lecture on techniques for survival in the woods, including what is safe to eat and what isn't. We are to "gather" our own food, but we **must not** steal from the local farmers.

The original groups again head out with the instructors posing different requirements to be met – including navigating. We have to rappel from a fairly high cliff; there is no net to catch anyone who falls. It looks very scary, but their instructions are excellent. No one falls while doing this training, however, one or two have some problems during the descent! Once I get the hang of it I think it is fun! We end up completing a big circle by the time we are back in camp in the evening.

We spend another night on the ground in our tent with a fire burning all night, but the air is not as cold as it had been the night before – thank goodness!

At dawn the instructors return and explain the "game-of-the-day" to us. We have to make it to a small "safe zone" about 5 miles away, without being captured by the "enemy" soldiers who will be patrolling the roads in vehicles and looking for us on foot over the terrain we must cross!

We are shown on a topographic map our destination and the boundaries of the exercise area. We have to stay within this area as we make our way to the buses, which are in the safe zone, during the allotted time. The buses will leave on schedule, not a minute later!

We have eaten almost nothing, because this time of year there is very little to scrounge – our only source of food. Then comes our highlight for the day – an AD-5Q from VAW-11 flies overhead and drops food for us!

(At this time, this mission is a big surprise to me because I did not know VAW-11 did this. I also did not know we had any 5Qs, because they are not parked with the 5Ws. This is a fun mission to fly for VAW-11; after returning from cruise I get to fly a couple of these flights. The Q-model is for electronic counter-measures; VAW-11 has only a few of these models.)

It is fun hunting and finding the small parcels that have scattered during the drop. We all feel a *lot* better with some good food in us!

It is time to start the evasion exercise by attempting to cross "enemy" territory to reach the buses within the allotted time. The first big problem is to split up so we will be less likely to be caught, so we leave a few at a time.

I have gone about a mile and a half by running and walking within the woods and crawling along in the low brush in the open spaces while trying to remain hidden from view. If one stayed in the woods without crossing open spaces it would have taken more time than was being allowed to reach the buses. I see a clump of large bushes, so I crawl inside to drink a little water, rest and check my bearings. Just as I am leaving, I hear, "Halt! Hands-up."

I turn around and see a marine crouching outside the bushes with a pistol pointed at me! His uniform is a nondescript "enemy-type" fatigue. Likewise, his gun is a foreign automatic rather than the Marine standard issue 45-caliber handgun.

No one has mentioned that the "enemy" is actually armed! It is all too real.

He gives me orders on how to come towards him without getting shot. He is playing his part to the hilt in a very convincing way. I know it will be difficult for you to understand, but I cannot trust this person not to have an "accident" – I am *scared*.

He marches me over to a paved road where I will be picked up and taken to the POW camp. He has me lie face down in the dirt with my hands clasped on the back of my head and tells me if I move he will kill me!

After a short while of not hearing him, I nervously peek under my arm and see that he has apparently gone to capture others. So I quickly scan my surroundings and listen for other activity. All is quiet, so I make a dash across the road and hide in some bushes. I then very carefully make my way to the buses without getting caught again. About 3/4ths of the "good guys" have arrived ahead of me.

Finally the allotted time has expired and the buses head to our base, from which we had started at the beginning of the training. There is a rumor that one or two have been left behind – they hadn't arrived within the time period and they hadn't been captured.

We pile out of the buses and walk into the main building. We are in the darkened front section of the building. On the stage, which is screened by a see-through curtain, is the prisoner interrogation room. We can see through the curtain, but the people on stage cannot see out. We are told to be quiet, listen and watch to learn how prisoners will be treated.

I won't tell you what we see and hear, but it is *very* real – the prisoners (our fellow trainees who have been captured) are nearly naked. They were kept in individual cages behind the building prior to being led in for interrogation, one at a time. I know it is all too real for them also!

I am so glad that I have used my head and escaped!

Then – surprise – we are given a bus ride back to NAS North Island. Our "beach" is that part of the airfield that is bordered by the bay and the ship channel into San Diego harbor (where the Navy sea planes operate). I had not noticed any people in this isolated area of the field while flying from North Island.

We settle into our new quarters – the sand dunes overlooking the bay.

Fortunately, we catch enough crabs for everyone within our group to have his hunger satisfied. We boil them, using seawater, in an old gasoline can. The other groups do likewise – we are all helping each other. The crab tastes great!

The next day is spent on how to survive within a beach environment. We have outdoor lectures covering all aspects of various types of beaches plus practical demonstrations and first hand experience of what we have learned with respect to our beach.

At night we have more good tasting crab for dinner.

There are some K-rations or some such on a small hill off in the distance. If we can make it to the hill without being caught, we can have the extra food and return without further hassle. Whether or not we participate is optional. If we had not had such good luck with our crabbing, I assume most everyone would have tried for the food.

I decide I'll give it a try – since there is nothing else to do for entertainment. I am about halfway there when another "survivor" joins me. Unfortunately, he makes more noise than he should and the two of us are caught. The guards are sailors rather than Marines, and though they have rifles there is no mistaking that this is *not* serious. The other prisoner is marched off.

I am left with two sailors who don't seem to know what to do with me. Since I have given my name, rank and serial number they know I am an officer. They think they will have some fun, and take advantage of the situation. So they order me to do push-ups. I say I have never done any push-ups and that they will have to show me how to do them. They get into a *big* argument over who is going to do the demonstration. Finally, one of them demonstrates to me how to do a push-up. So I do a few push-ups. Then they ordered me to do some other exercise that I claim I don't know how to do. Again it has to be demonstrated to me. After the third repetition of having to demonstrate the exercise to me, the demonstrator has had enough, and he asks, "What calisthenics do officers do?" I reply, "Officers never do any calisthenics and I never did any in college either." They walk away mumbling some derogatory remarks about officers – and I have had my entertainment!

The next morning we say good-bye to our beachfront "quarters" and walk about half the width of the airfield along the channel to a low-lying concrete building. Within this building we have lectures and demonstrations on how to survive on the ocean. Including the proper deployment of our one-man-life-raft and use of the emergency equipment packed within our seat pack. This includes the best way to signal with a signaling mirror. (After this I always carry an extra signaling mirror and a packet of fishhooks in my flight suit.)

The parachute is on the bottom of the seat pack – which we sit on in the aircraft. They review the proper procedures for steering a parachute during a descent, how to prepare for landing in the water, the proper way to enter the water and then the best way to quickly collapse a parachute, if required.

Early afternoon they demonstrate how to be picked out of the water by a helicopter. We then strip and store our shoes, jacket and flight suit in a locker and are loaned another flight suit. Barefooted we climb into a cattle-car type tractor-trailer "bus" that takes us to a different section of the beach to practice what we have learned.

We are divided into groups of about eight. A whaleboat takes one group at a time out to the middle of the channel, where one jumps into the water, is retrieved by the helicopter and then flown back to another section of the beach than the area where the others are waiting to go out in the boat. The way they have explained it, I think it will be easy to be rescued. From the shore and later from the boat, it doesn't look as though anyone is having any serious problems.

In the boat when it is my turn, I jump into the water and the boat pulls away. I tread water as the helicopter comes towards me while lowering the "horse-collar." As the helicopter decreases its altitude during the final approach overhead, surprisingly a *heavy* spray shots up from the water surface for at least 2 feet and suddenly engulfs me! It seems that each breath is about 50% water in the form of small droplets. From the shore and whaleboat this helicopter-induced spray has appeared rather innocuous, but it is suffocating!

The instructor has informed us how to put the horse-collar over our head and shoulders. He also demonstrated the *wrong* way – and said that only a *very* few would do it correctly in the water the first time.

During the lecture the right way seems rather straightforward and reasonable. However, we had not been able to practice it out of the water. Sure enough, much to my chagrin, I start putting it on the quickest way – which is the wrong way - because I want to get out of all this smothering spray as soon as possible, before I drown!

When the air crewman standing in the doorway of the helicopter begins yelling at me I cannot understand what he is saying, but I know immediately why he is yelling. I instantly stop putting the collar on the wrong way and take the extra step required to do it correctly. Then up I go, either by the winch, the helicopter rising a few feet, or both. Above the spray I am finally able to get a good breathe of real air!

The heaviness of the spray has caught everyone by surprise! Why this was not included in the presentation is a mystery to us all. I think this is the main reason most students failed to get into the horse-collar correctly the first time!

This is my first, and my last, trip in a helicopter. I cannot believe how much the body of the craft shakes during flight.

Now that we are all wet it is obvious why the cattle-type bus has been used for transport. We are delivered back to the building where we have stored our clothes. Upon dressing we are all bused back to our Monday morning starting locations.

I am amazed how much useful information and training has been provided in only five days. It certainly instilled confidence in me during my flying, that I can handle an emergency on water or land.

* * * *

Based upon what I have read and learned directly from talking with a former Vietnam POW (who was a pilot I had known in VAW-11), the simulated prisoner interrogation was very realistic! When my flying days were over, I was also thankful that I had not had to make use of this information!

9
CARRIER OPERATIONS & DEPLOYMENT

Background

USS *Hornet* (CV-12) (US Navy Official)

USS *Hornet* (CV-12) was commissioned on 29 November 1943 as an *Essex* class carrier with a straight wooden flight deck. She was mothballed in January 1947. She was pulled from the reserve fleet in March 1951 to undergo extensive modernization (SCB-27A), mainly to allow her to handle jet aircraft – including upgrading her catapults to H-8 and the arresting gear engines to Mk 5. She was recommissioned at the New York Navy Yard on 11 September 1951, and reclassified as an attack carrier (CVA-12).

She was again extensively modernized (SCB-125) in the Puget Sound Navy Shipyard from 28 January till 17 August 1956. The changes included the enclosed hurricane bow and the installation of an angled deck (12 degrees off-center) for landings, a British innovation. The wood in the landing area of the angled deck was replaced with steel and the number of arresting wires was reduced from about 11 to 5. In addition, the Mirror Landing System (MLS) was installed, another British development. Unfortunately, her two hydraulic catapults were not upgraded to steam powered, as were many of her sister ships that had gone through modernization.

Both catapult tracks are the same length, but they are staggered; the starboard track terminates at the end of the flight deck, while the port track terminates several feet short of the deck-edge - as shown on the previous page.

Since the MLS is relatively new, AD pilots, at least those flying in the Pacific area, are required to qualify on *both* paddles and mirror. Thus, if the mirror system fails there is a backup method for recovering the Skyraiders. The AD day-and-night carrier qualifications (CarQuals) are a minimum of five day and five night arrested landings with paddles, five day and five night arrested landings with the mirror plus three catapult shots, at least one of which will be at night. A carrier landing on an angled deck that is not arrested is called a bolter, and counts as a landing, but not as an arrested landing.

Prior to carrier qualifications, all pilots who have not qualified in-type or have not made a carrier landing within the last six months have to complete Field Carrier Landing Practice (FCLP). This is accomplished by simulating carrier approaches at a remote runway with a Landing Signal Officer (LSO).

A carrier landing is accomplished under the supervision of a LSO, who stands on his platform well aft on the port side of the ship, facing aft. During the final approach the LSO can signal up to 14 desired corrections with his arms/paddles and legs plus 2 *mandatory* commands with his arms/paddles: "cut" and "wave-off."

With the mirror, all communications (except the "cut" and "wave-off") to the pilot from the LSO must come over the radio. A pilot never talks to the LSO during an approach, except to make specific reports. At any time on final that the LSO thinks the landing cannot be safely made, he signals a "wave-off."

During final approach, a propeller-driven aircraft with a tail wheel is flown at nearly the 3-point attitude (the bottom of all wheels at the same level) at just above the stall speed. At the cut-signal, the pilot performs the following sequential actions: moves the throttle to idle, eases the stick forward to prevent the plane from stalling and crashing into the fantail of the ship, then eases the stick back to decrease the rate of descent and stall the plane just inches off the deck for the 3-point landing.

Instantly when arrestment is felt the stick must be pulled *all* the way back to keep the nose from dropping too far and having the propeller strike the deck during the run-out of the arresting cable. During arrestment, the tail wheel will not stay on the deck, even if it makes it there during the initial touch down. An AD is usually level to slightly nose down during most of the run out of the cable. However, the AD-5W drops its nose more than other ADs during arrestment.

CarQuals

At 0800 on Monday, 23 September 1957, we are waiting in our respective ready rooms aboard USS *Hornet* (CVA-12) as she steams off the coast of southern California. It is the first of three days of day-and-night carrier qualifications and re-qualifications for the Douglas Skyraider pilots who will deploy on this ship as part of Air Task Group 4 (ATG4) in January 1958. The Skyraiders that are spotted on the flight deck have the designations of AD-6 and AD-7 of squadron VA-216, AD-5W of Detachment Kilo of squadron VAW-11 and AD-5N of Detachment Kilo of squadron VA(AW)-35. All these planes were loaded aboard at dock side.

During pre-flight and flight operations aboard ship, the Air Boss and his assistants are in "PRI-FLY" spaces on the upper port side of the island - this is the "airport control tower" of the ship.

In the ready room we hear the command from PRI-FLY, "Pilots, man your planes!" We quickly climb the ladders to gain access to the flight deck via the island. Stepping onto the deck we turn left and see the planes closely packed on the aft end of the deck. We have been told that our first "takeoff" will be via the starboard catapult, with subsequent deck launches between arrested landings. The goal this day is to finish the paddle-controlled day arrested landing requirements for all pilots.

Our four AD-5W aircraft are furthest aft on the flight deck.

Having completed our check-off list down to starting the engine, we sit in our planes waiting for PRI-FLY to announce, "Pilots, start your engines!"

Upon receiving this order we all start our engines and complete the takeoff check list while we wait for the cylinder head temperature to rise and the oil pressure to drop until they are within their normal operational ranges (in-the-green).

Then, one by one, the planes taxi forward out of the pack under the direction of a plane director. As soon as a plane is clear of other aircraft, the director signals the pilot to unfold his wings while taxiing forward towards the bow. Each plane is past to another director further forward on the flight deck. To the extent that I can, I watch the planes being catapulted while I also respond to signals. I see that shortly after each plane climbs away from the bow, a small dark object falls from the tail of the aircraft. I wonder what these objects might be.

Finally I am turned over to the plane director of the catapult crew and follow his directions. Soon my plane is aligned with the 220-foot long catapult track and I stop with the tail of my plane over the start of the track. This puts the end of the flight deck (and track) 4 1/2 plane lengths (180 feet) in front of me. A catapult shuttle rides on and between two steel rails, which are closely spaced and flush with the flight deck. A bridle (cables) connects the shuttle to two open hooks on the bottom of the AD wing, one on each side of the fuselage. The tail of the aircraft is secured to the deck with a bar connected to the plane by a holdback ring that prevents the plane from moving under full power; but the holdback ring breaks when the cat is fired. When the shuttle stops at the end of the track, the cable-ends slip out the back of the open hooks as the AD flies free of the ship.

After a plane is connected to the shuttle, tension is applied to the cables that compresses the oleo struts of the landing gear and firmly holds a plane to the deck of the carrier during the catapult launch. This is accomplished with an AD by moving the shuttle slightly forward after the tail is secured. My plane is "tensioned-out" and the catapult officer signals me to go to full power. I advance the throttle to full power and wrap the fingers of my left hand around the top of the throttle *and* "static-bar," and place my thumb below the throttle. I had folded down the static-bar after I had started my engine, in preparation for my first cat shot. Gripping the static-bar and throttle together prevents me from retarding the throttle during the shot. If this should happen, my plane would "land" in the water.

Catapult Officer signaling to launch Skyraider from starboard catapult of *Essex* class carrier. (US Navy Official)

I check my engine gauges and flight instruments, and salute with my right hand to indicate everything checks out and I am ready to be "shot." I then place my right hand on the stick, thumb on left side and fingers on the right side, with thumb and index finger level to form a "U" around the back side of the stick and firmly lock my elbow against my right hipbone while holding the stick in a "neutral" position. This hand and arm positioning should prevent the stick from moving back because of the force of the shot, which if this occurred would result in a stalled "landing" on the water. I place my head firmly against the headrest and focus my eyes on the gyro horizon, as we have been instructed to do by our squadron LSO.

If you observe this sequence of events from outside the cockpit, there is very little time between the salute of the pilot and the firing of the catapult by the shooter. But I experience this time in slow motion, as I was to do in all future cat shots. It seems like forever between the time I have saluted and the time the catapult is fired.

Hydraulic cats only accelerate a plane over about the first third of the track. Thus, the acceleration on a hydraulic-powered cat is about 3 times more than on a steam-powered cat. In addition to the hard, sharp kick in the back one gets from a hydraulic cat, the associated "blinding" acceleration forces almost *all* of the peripheral vision to be lost, like instantaneous *severe* tunnel vision.

The shot is *much* harder than I have anticipated and the loss of peripheral vision is virtually instantaneous. I do not see the bow of the carrier go by. In fact, I cannot see anything outside of a circle of about 5 inches in diameter, centered on the gyro horizon instrument.

I immediately raise my gear with my left hand, without seeing the gear lever in my peripheral vision. As I shift my eyes from the gyro horizon to the actual horizon in my windscreen I see too much water and not enough sky. I instinctively begin pulling back on the stick with my right hand, but the resisting force on the stick is too much for me to stop the loss of altitude. I add my left hand and arm to the all-out pull on the stick, but I still do not have enough strength to prevent the loss of altitude. Instantly I put my right thumb on the elevator-aileron trim tab button on top of the stick and pull it back and hold it, until the nose begins coming up and the flight pressure on the stick is back to normal.

The flight deck is only 60 feet above water.

I knew immediately that I had not held low enough on the stick. I had expected that with my elbow braced against my hipbone this would be enough to prevent the stick from moving aft during the shot. Though the stick had just barely moved rearward, it was just enough for the web of my glove to engage the elevator-aileron trim tab and put in "down trim" until I was airborne.

After the flight, Darrel Westbrook tells me, "I was following you on the cat and it really scared me when I saw you drop out of sight for a long time! I thought you had gone in." The LSO tells me, "The officers and men on the bridge thought you were going in when they lost sight of you. When they finally did see you again you are making a large wake in the water with your prop wash."

While climbing out I place the canopy control to "close", reduce power (throttle then propeller control) to normal climb settings, milk the flaps up, move the mixture control from "rich" to "normal" and join the landing circuit at 260 feet. (Please see the Appendix for more technical information.)

I start turning left onto the down-wind leg when the plane ahead of me is on his down-wind leg. Once on the down-wind leg I start my landing check-off list, which includes opening the canopy, lowering my tailhook, flaps and landing gear, and etc.

At the 180-degree position (opposite the stern of the ship) I identify myself with my call sign, Romeo Romeo and the last two digits of my plane's number painted on its cowl – Romeo Romeo comes from the squadron's identification letters, two R's on the fin-rudder of VAW-11 aircraft. This information is passed to the arresting gear personnel to ensure the tension of the arresting cables is set for the recovery of an AD-5W. After saying, "Romeo Romeo two five, flaps, gear, and hook, down." I ease off the power and start a left hand descending approach to the stern of the ship at about 110 kts (127 mph). Just past the 90-degree position I am established at 90 kts (104 mph) in a 3-point attitude ready for my final approach. I pick up the LSO around the 45-degree position and follow his signals to a "cut" and my first AD arrested landing.

The LSO stands just off the edge of the flight deck on the port side of the ship – he is looking towards you in following picture. Further up the deck there are five arresting wires. Off the port side of the flight deck, just a little forward of where the angle deck bends back towards the centerline of the ship, is the Mirror Landing System (MLS), which is gyro-stabilized. The rectangle delineated at the end of the angel deck is the #2 elevator. The superstructure on the starboard side of the ship is called the island.

In the grove for landing aboard USS *Hornet* CVA-12. (US Navy)

Brakes are not used during an arrestment. The wind across the deck starts the plane moving backwards as soon as the plane is fully arrested (which is *very* rapid). This backwards motion is desired for a very short distance in order to clear the arresting cable from the tailhook. A plane director signals the pilot to stop the plane as soon as the plane has rolled backwards far enough for the cable to disengage itself from the tailhook (or allow an arresting gear person to manually clear the cable from the tailhook).

The plane director then signals the pilot to raise the tailhook and start taxiing forward and to the right to clear the landing area as soon as possible, because another plane is making its final approach. It is consider very bad etiquette to force the plane behind you to get a wave-off because of a "fouled-deck."

I taxi under direction to the right and then to the left and am aligned with the imaginary centerline of the straight part of the flight deck for the scheduled deck launch. But I am not signaled to stop; instead I see the signal to raise my flaps, followed by the fold wings signal while I am taxiing.

When my plane is centered on the forward elevator, I cut the engine, as signaled. The elevator drops quickly to the hangar deck; to my surprise I see a lot of ADs. I am confused. I thought they had put me on the elevator because they hadn't liked my "takeoff" and were not going to let me fly until the LSO had a talk with me.

But why were all these other ADs down here?

My plane is pushed to the place where they want it and is secured to the deck. I climb out of my plane and as I step onto the deck, our Maintenance Chief informs me that my tail wheel is "shot" and has to be replaced - along with *all* the other AD tail wheels.

There is a little projection in the starboard catapult track, that no one had seen, that has destroyed the hard rubber part of each tail wheel during the catapults. No one who is in authority either saw or comprehended what was happening shortly after each AD was climbing away from the ship, as some small dark object was falling from each one.

Since the ship only has the ADs aboard, all flight operations are brought to an unexpected early termination, as all 21 ADs are "Down, awaiting part" - a new tail wheel.

Since the ship's supply locker does not have 21 AD tail wheels, the ship makes an emergency order for a COD (carrier onboard delivery) aircraft to deliver the required tail wheels to the ship from NAS North Island.

Later in the afternoon after our tail wheels have been replaced, I complete 11 deck launches and arrested landings, with the LSO using paddles.

During the afternoon of the second day we have a short warm-up to better prepare us for our night carrier landings. I complete one cat shot and two arrested landings with the LSO using the mirror landing system (MLS).

Over an hour before scheduled launch time all the illumination in the ready room is switched from white to red so our eyes will be adapted to the darkness as we man our planes on the flight deck.

Since we need at least one night catapult shot for our quals, our initial departure this night will be with the assistance of the catapult. We are informed that the overcast is lower than they would like for carrier qualifications, but they think it will be okay for paddle-type approaches - at least they are going to give it a try. Of course the "they" will not be flying.

Soon the order from PRI-FLY is announced over the 1-MC (loudspeaker), "Pilots, man your planes!"

As I step out of the ready room into the passageway, I note that the passageway lighting has also been switched from white to red. Stepping onto the flight deck, I see no aircraft!

It is a very dark night below the low overcast.

When I get to where the planes are I must be very careful not to trip over the tie-downs, because I cannot see them. Theoretically, we each know where our individual planes are spotted on the deck, but in the darkness I must ask the plane captain to ensure that I have found my plane because I cannot read the large black aircraft numbers on the light gray engine cowls.

We all start our engines without difficulty after PRI_FLY announces, "Pilots, start your engines!"

The trip to the catapult is like being at a magic show. The glowing wands used by the plane directors appear to be suspended in mid-air with ghostly faces appearing and disappearing, depending on the position of the wands.

As I taxi towards the starboard catapult, I notice that the forward edge of the flight deck is delineated with a row of low intensity white lights and that the second light from the right is burned out, leaving a noticeable gap in the spacing of the lights.

Then I begin wondering again how it will be to commence flying sixty feet in the air, as my peripheral vision is returning, and being totally dependent upon my flight instruments - for there is no horizon to see on this dark night.

Much to my relief, the immediate instrument flying following the cat shot is going very well, but as I am climbing out I abruptly enter the clouds and lose contact with the wing and tail lights of the plane that is ahead of me. I ease off the power and lower the nose to maintain a constant airspeed while I descend to 200 feet.

I still cannot see any aircraft lights ahead of me or off to my left, but it is time for me to commence my standard rate left turn to fly the up-wind turn of the circuit. I roll out of the turn when my heading is opposite to the out-bound heading, which puts me on the down-wind leg of the circuit.

I can see no lights anywhere!

Based upon time of flight when I should have been at the 180-degree position (opposite the stern of the carrier) I still see no lights. I ease off some power and start a very gradual descent, while maintaining my heading.

As I am wondering what I should do next, I am down to 150 feet; I see the running lights of a ship off to my left. I continue straight ahead until the ship is off my left wing tip and turn to come up behind the ship so I can get a look at it. As I fly over the ship I notice the spacing between the running lights look as though they could belong to a destroyer. So hoping I have found the plane guard ship, rather than a commercial fishing boat or cargo ship, I take up the heading the ship is steaming, which is the same as the carrier's. (A plane guard ship is a destroyer that trails the aircraft carrier at night to pick up pilots that "land" in the water.)

Sure enough I shortly see the lights of the carrier and flying planes. I rejoin the circuit at 180 feet and make my up-wind turn when the plane ahead of me is off my left wing tip - this establishes the time-interval (about 90 seconds) the Flight Operations Officer wants between recovering aircraft.

Supposedly, 260 feet (200 feet above the flight deck) is the *minimum* altitude for paddle-controlled night carrier qualifications. So much for the "book."

At the 180-degree position I identify myself with my call sign and "flaps, gear, hook, down," and ease off power as I start a left hand descending approach to the stern light of the ship by flying on instruments, with an occasional glance in the direction of the ship.

I pick up the lights of the LSO's flight suite and paddles (at about the 45-degree position – the same position as in day time) and follow his signals while also periodically scanning my flight instruments.

As I get closer to the ship the centerline lights of the angled deck suddenly appear. One can see these lights through a very narrow window, in both azimuth and elevation.

At the "cut," as I lower my nose, the centerline lights are gone and I land in a sea of black.

The port and starboard edges of the flight deck are not outlined in lights. The jerk of the arresting gear feels like heaven to me!

All hand (wand) signals to an AD-5 are given from in front of the plane, but off to the left side (roughly the 10:30 position), the side where the pilot sits. In pre-carrier training, we were told to *instantly* follow all directions of the flight deck crew, because it is choreographed to the nth detail, and their life and your life depend upon complete obedience - doubly so at night.

At the sign of the "X" from the wands of the plane director I apply my brakes to stop the backward motion of the plane, raise my tailhook as signaled and start taxiing, following the signals of the wands for a deck launch. After almost no taxi, I am surprised to be passed to a launch director.

He instantly gives me a stop signal and then the power-up signal. After a quick glance at my instruments while I increase the power to the pre-launch setting, I look at him and nod my head that I am ready to go.

He immediately gives me the signal to launch.

I look forward through the windscreen for the first time since my landing as I release my brakes and start advancing the throttle to full power. As my plane's tail lifts and its nose goes down, I can finally see in front of my airplane. I am very surprised how close the end of the flight deck looks at night and that the second light from the right is *not* out.

In my peripheral vision to the right I notice another set of dim white lights, with the second from the right light being out. I quickly conclude that: I had never been taxied clear of the landing area of the deck, I had been mistakenly aligned with the angled deck rather than the straight deck, and there is no way I can get to full power and flying speed before reaching the end of the angled deck.

So I *stand* on the right rudder pedal to head for the center of the bow lights. After I have completed over half my turn towards the bow, PRI-FLY announces on the bullhorn/radio (?), "Right rudder, right rudder!"

So now I know that my conclusions were correct and that I am taking the right corrective action!

Once I am heading for the bow, I finish adding power for my takeoff and complete my first night deck launch.

This is the first and last time I ever had to make a turn in the middle of a takeoff run in any airplane. I do not know how close my left wheel came to going off the deck. Later I am told that some of the flight deck personnel were undergoing training – which probably explains the screw-up.

This night I complete one catapult shot, four deck launches and my required five arrested carrier landings with the LSO using paddles.

During daylight hours of day three I obtain three deck launches and arrested landings, with the LSO using the mirror - which completes my five-day mirror arrested landings.

At night I complete one catapult launch, five deck-launches and six arrested landings, with the LSO using the mirror - all in 30 minutes. This implies we have gotten our landing interval down to about 60 seconds. I still have no bolters.

I enjoyed deck launches (excluding my first night deck launch), and preferred paddles to the mirror for both day and night AD carrier landings because of the flatter approach of the paddles with the associated higher power setting than the steeper decent of the mirror approach with its lower power setting. Thus the transition to the 3-point, full stall landing was easier for me with the flatter paddles-approach.

However, I thought the MLS would be great for planes with tricycle landing gears! Shortly after this exercise COMNAVAIRPAC put a stop to all paddle controlled landings for ships that had a mirror. So I considered myself very lucky to have been able to make the transitions from a straight-deck carrier to an angled-deck carrier and from paddles to the mirror. These were my last paddle-controlled landings.

I always hated hydraulic cat shots! Compared to hydraulically operated catapults, steam powered catapults are a *great* improvement (as I found out about a year later) - but not as much fun as deck launches!

Scared

VAW-11 Detachment Kilo is on another pre-deployment fleet exercise, this time back aboard USS *Hornet* with Air Task Group 4 (ATG4). Most of the ADs of ATG4 are spotted on the aft portion of the flight deck. We are in our assigned ready room – a jet squadron ready room. Since we are a detachment we share this ready room, actually they share it with us – since it is theirs. In this exercise we are launching only one aircraft. However, even when only one aircraft is to fly a mission we man two aircraft, because planes do go "down" during the checks made before launch. So two crews (six people) are waiting for the command to man our planes, when we think we hear AD engines starting!

Shortly afterwards our ready room loudspeaker barks, "Pilots, man your planes."

When we step out of the island onto the flight deck we look aft – all the ADs, except our planes (which are spotted in the rear), have their engines running. The planes have their wings folded over the cockpits, and are almost touching at their stubby wing edges. The planes are from deck edge to deck edge.

There is only one quick way to our planes – walking between adjacent whirling propellers while ducking under the wing stubs, not losing our balance as the deck heaves and rolls under foot, and not tripping over the stubby wing tie-downs and falling within the propeller arcs. It is really loud, unnerving, and dangerous – flight-deck crews and plane crews did *not* do what we are being required to do! The flight deck crew's faces show real concern for our well being!

CVA-12 flight deck with ADs ready for deck launch. (US Navy)

What had happened? The personnel working for the Air Boss had forgotten that our detachment is in a jet squadron ready room – therefore the announcement to man your planes (which is just for AD crews) was not sent to our ready room until after the Air Boss noticed two ADs with stationary propellers.

Fortunately, this was the only time I ever had to do anything similar to this.

47,000th Arrested Landing Aboard Hornet (CVA-12)

Traditional cake cutting ceremony for a thousandth landing aboard an aircraft carrier is pictured. Commanding Officer, Captain T. F. Connolly USN, is extending congratulations to LTJG Harry D. Hamilton USN of Carrier Airborne Early Warning Squadron Eleven (VAW-11) Detachment KILO.

I am lucky enough to make two more thousandth arrested landings on other *Essex* class carriers in the Pacific.

Pilots to their Ready Rooms – On the Double!

It is about 0300; Darrel Westbrook and I are fast asleep in our stateroom during a two-carrier exercise well offshore of San Diego in October 1957. Suddenly, we are awakened by the announcement: "Now hear this! Now hear this! All pilots, dress, report to your ready rooms – on the double!"

Darrel and I spring out of our racks, jump into clothes and dash to our ready room.

After *Hornet's* last recovery for the day, the aircraft handlers spotted the flight deck for the first launch of the morning. All Skyraiders in the morning launch, which includes two of our planes, are on the aft end of the flight deck – where planes land. The jets are spotted further forward, some of which are also in the way of recovering aircraft.

In our ready room we are informed that a S2F has crashed during landing on the other carrier in this exercise. Their flight deck is afire and there are more planes aloft which must land. If this ASW carrier cannot extinguish the fire and continue recovering aircraft, we are to man our planes and taxi them forward, as directed, to clear the landing area for use by the planes still airborne. A flight deck can be most quickly cleared for landing when aircraft are taxied, rather than pulled, off the angled portion (landing area) of the flight deck.

With much concern for our fellow airmen, pilots wait in their respective ready rooms for word to man planes.

Finally - it seems like forever, considering the situation – we hear the other carrier's message: "Our fire is out. We will be able to recover all our planes without assistance. Thanks for standing by!"

This accident and associated reaction makes clear that though carrier aviation is a dangerous profession – we always pitch in to help one another!

Hitch a Ride

After we complete the above exercise it is time to fly our 3 planes back to base. I have forgotten how/why I ended up without an airplane, probably lost a coin toss. But anyway, I can either stay with the ship as it returns to the pier at North Island or fly back with Darrel Westbrook. I want to go home, so I volunteer to be a passenger in the right seat, where the Air Controller normally sits. I am not looking forward to the cat shot – without me flying; but I want to get home.

Next morning, Friday, 11 October 1957, after the brief for the flight home and the announcement: "Pilots, man your planes" I walk out with Darrel to our plane, spotted on the starboard catapult. I am nervous, I must admit – I trust Darrel – but not as much as I trust myself!

He goes through all the checks and we are just about to be launched, when all of a sudden, it dawns on me that I will not be *responsible* if something goes wrong! For some reason this obvious realization just sweeps over me – and I completely relax!

I wonder how much of the tension of flying from a carrier revolves around being responsible! It is so easy to kill someone on deck or in your plane if you screw-up!

Maximum Endurance/Range

The airborne early warning (AEW) missions we fly in support of the fleet requires us to mainly operate our aircraft in two, modes - "maximum endurance" and "maximum range." These two modes refer to our power settings – rpm and manifold pressure. The rpm of the engine is determined by the pitch of the propeller which is altered by the prop-control and the manifold pressure is controlled by the throttle. For any given weight, there is a correct air speed based upon lift and drag that will allow the engine to consume the least amount of fuel (maximum endurance) or fly the greatest number of miles (maximum range). These performance curves are in the appendix of our AD-5W Flight Handbook. Unfortunately this section of our pilot's manual is marked "CONFIDENTIAL."

However, in order for us to obtain maximum performance from our aircraft we need this information in a quick and easy format while we are flying. So I tabulate this data for incremental weights of our aircraft from the curves and then type four cards to fit our kneeboard (one for each pilot of our detachment) on the rpm/mp versus weight for endurance and range. I do not mark these cards as confidential because I do not title the cards nor explain what the numbers in the columns are/mean. Our pilots, LCDR Bodger (O-in-C), Darrel and Mitch are happy to have the cards.

Unfortunately, some of our other detachments which are in training hear of these cards and also want them, then the Operations Officer, a Commander, hears of this and calls me into his office. He is upset that we have unmarked confidential data cards and wants them all destroyed. After heated discussions with other senior officers of the squadron, including prior O-in-C's of other detachments, I finally obtain permission for us (and others) to keep these cards – but they must be marked "CONFIDENTIAL."

Prior to this, the deployed detachments were using "rule-of-thumb" approximations for the prop/power settings – and were not differentiating between range and endurance.

Again, I think my aeronautical engineering background helped me to realize the importance of using this data operationally and putting it in a useful format for the pilots.

Deployment

USS *Hornet* (CVA-12) Passes under Golden Gate Bridge while leaving San Francisco Bay on 6 January 1958 for our WESTPAC cruise.
(US Navy)

(Note size of the North American AJ Savage Heavy Attack plane in front of ship's island.)

VAW-11 Detachment KILO 1958

COMNAVAIRPAC ORI SNAFU

My crew, an Air Controller and an Aviation Electronics Technician, and I are anxiously waiting in our ready room for the start of the Operational Readiness Inspection (ORI) by COMNAVAIRPAC personnel of USS *Hornet* (CVA-12) and her embarked air group, Air Task Group Four (ATG4). This consists of around the clock flight operations for several days in Hawaiian waters during January 1958. We are in VAW-11 Detachment Kilo, flying the single engine, propeller-driven AD-5W, a version of the Douglas Skyraider that provides Airborne Early Warning (AEW) for the fleet.

In response to the order: "Pilots, man your planes!" we dash into the passageway and up the ladders to the flight deck. We know our plane is on the port catapult and will be the first plane launched, in accordance with the AEW requirements of the ORI.

Stepping onto the flight deck, I see a sea of jet aircraft. Looking aft, all the Skyraiders, except ours, may be seen in the rear of the pack. Turning towards the bow I see an AJ Savage of VAH-16 Detachment Kilo on the starboard catapult and their other AJ spotted forward on the port side. The North American AJ is a large (will not fit on a flight deck elevator), three-engine (two reciprocating engines on a non-folding wing and a jet engine in the tail), heavy (about 65,000 pounds), attack.

Getting nearer to the bow I can finally see our plane on the port cat, directly in front of the second AJ. I am concerned for our welfare; the positioning of this AJ implies it will follow us on the cat.

USS *Hornet* (CVA-12) on 1958 WESTPAC Cruise (US Navy)

(Note ADs on fantail, AJ Savages in front of island and behind starboard elevator – elevator #3)

After preflight, I climb into my plane, which weighs about 15,000 pounds, and complete the takeoff checklist down to the starting of the engine, while the catapult crew finishes their attachments. Soon the flight deck bullhorn blares, "Propeller aircraft - start your engines!"

As my engine becomes warm enough to go to full power, the ship starts turning into the wind and increasing her speed to produce 30 kts of wind down the deck to commence flight operations. Responding to the Catapult Officer's signals my flaps are down and I have gone to full power. I am relieved! This normal sequence of events means the cat's hydraulic pressure is not set for launching an AJ.

Surprisingly, before I finish checking gauges for normal values, the Catapult Officer signals "Power Back." I pull the throttle back to idle. Shortly he signals "Flaps Up." I raise my flaps and wonder if they have seen something wrong with my plane.

As the ship starts turning out of the wind and slowing down, I think the launch must have been delayed; hence the unexpected signals. After a 180-degree turn, she continues slowing to a speed just adequate to hold a downwind heading. During this maneuvering the wind across the deck has changed from a 30 kt headwind to at least a 10 kt tailwind.

Finally, I realize this unusual maneuvering has a purpose: to launch our plane flaps-up with a tailwind using the force required to catapult the AJ waiting behind us! I wonder if we are expendable. Just before the full power signal, I warn via intercom, "This is going to be the cat shot from *HELL*"

I push the throttle all the way forward, check gauges, and salute the Catapult Officer. The salute indicates we are "good to go." He responds with a signal to the "Shooter" to launch us.

A hydraulic-powered cat provides acceleration to the plane for only one third of the distance of a steam-powered cat. Therefore, the force is 3 times greater than that of a comparable steam-powered cat shot. Because of the extreme acceleration during a *normal* hydraulic-powered cat shot peripheral vision is instantly lost. This tunnel vision is so extreme that following a pilot's salute he must center his vision on the gyro horizon, because little else will be seen. With the greatly excessive force of *this* launch, we are enduring a force over 400% greater than normally applied to an AD-5W; I instantly lose *all* my vision!

It is an *absolute* rule for a safe launch: the plane being catapulted must not become airborne before reaching the end of the track. The "flaps-up with a tailwind" solution is giving us the best *chance* of obeying this rule, because it minimizes the plane's lift during the shot. If only our plane holds together!

This dire situation has been created by the AEW requirements of the ORI because, evidently, COMNAVAIRPAC *assumed* the *Hornet's* catapults had been upgraded to steam power, when she was modernized after recently being pulled from the mothball fleet of World War II carriers. Unlike steam-powered catapults, the pressure of a hydraulic system cannot be quickly increased after it has been decreased. Therefore, the pressure of the cat could *not* be lowered for our shot and then increased for the AJ in the time allowed in the ORI to launch the first wave of aircraft! The SOP of a carrier with hydraulic-powered catapults is to *always* launch aircraft in descending order of hydraulic pressure.

My sightless condition persists for a short period after we become airborne. I hold the controls in neutral positions, which keeps us out of the water. As my vision returns to normal, I call my crew on the intercom to check on their condition and ask them to check for damages. We are all aching from the shot, but surprisingly there is no apparent damage.

It is assumed appropriate modifications were made to the AEW requirements, because during the remainder of the ORI all AD-5Ws are launched with the other Skyraiders, in accordance with their weight.

Based upon available evidence, no other operational Skyraider has undergone a cat shot with this much excessive force!

The second morning of the ORI starts off badly, LCDR McGinnis, O-in-C of VA(AW)-35 Detachment Kilo, went in (crashed) while performing a dive-bombing run through clouds in an AD-5N. All three members of the crew are killed. VA(AW)-35 is our sister squadron at NAS North Island.

I fly a late morning AEW flight that starts with a hydraulic catapult shot and ends 3.8 hours later with an arrested landing. Arriving in the ready room I see they are preparing for the next launch. Darrel Westbrook, who is the Operations Officer of our detachment, cannot find another of our pilots, so he schedules me to man the stand-by plane for the next launch.

I tell him I am really too tired to fly another mission so soon. He says, no problem, he is flying this flight and all I have to do is start and taxi my plane as directed. So in response to: "Pilots, man your planes!" I scramble to the flight deck with a different crew than had flown with me on the last mission.

After the preflight, I climb aboard and follow the subsequent order for propeller aircraft to start their engines. I spot Darrel two plane columns over, at the starboard edge of the flight deck. His engine is running, a good first sign!

The aircraft directors finally work their way back to the Skyraiders. I look over and see a director signaling for Darrel to start moving his plane forward and then a signal for a slight left turn to get his plane out of the pack. Suddenly, I see his plane is too nose-high and too tail-low! (This incident is documented in our cruise book.)

When the director starts me, the backup plane, taxiing I know what has happened – Darrel has taxied his tail wheel off the flight deck!

So it is another cat shot and AEW mission for us. After 1.8 hours of daylight we are flying at night with some actual instrument time during the rest of the flight.

Upon returning to the ship, my signal is Charlie, so I enter the landing circuit. During the final approach, when we get within the cone-of-view of the MLS (Mirror Landing System) I see the "meatball" and report to the LSO, "Romeo Romeo two seven has the ball." Just before the ramp, I get the "cut" signal from the LSO via the mirror – flashing row of green lights - and proceed with a normal night carrier landing. We land three-point on the darkened deck and head straight down the canted deck. I slowly begin easing the stick to the rear in preparation of pulling it all the way back when our hook engages a wire.

The end of the angled deck is rapidly approaching!

Finally I realize we are *not* going to be arrested – this is going to be a bolter. So I begin moving the throttle forward at a good rate. But if I move it too fast, my plane will torque roll off the end of the deck and crash nose down and inverted into the black ocean below. However, if I move it too slowly, we will not have reached flying speed by the time we run out of deck; and will dribble off into the water nose first. This is when a carrier pilot *must* have a good feel for his propeller-driven airplane!

Likewise, if a pilot jams on the throttle after a way-off, he will also torque roll into the water or the ship! I am referring to our high-powered propeller-driven aircraft, not jet powered aircraft in both these incidents. Please see next page for sequential pictures of an AD-1 in a torque roll.

We fly off the end of the angled-deck and rejoin the landing circuit for another try. This time there are no problems! This happened after flying about 7.4 hours this day – and on a flight I wasn't suppose to fly.

AD Torque-Roll Following Wave-Off by LSO (US Navy)

This was my first and last bolter!

A bolter may or may not be the pilot's fault. With a ship rolling and pitching it is easy to miss a wire or two. Also the tailhook can and will skip a wire or even double or triple skip wires.

I guess this is as good a place as any for me to mention how proud I am of our Air Controllers, who ride in the right seat next to me, and Aviation Electronics Technicians, who ride in a separately enclosed area behind me – they *never* panic and break the rule of no talking during takeoffs and landings!

By the way, we get to fly from the ship to Ford Island at Pearl Harbor prior to the ORI. Don't remember who else besides our detachment got to do this, but I do know it is only Skyraiders. The runways are not very long. We fly some flights from Ford Island while our ship is tied up along side a pier of Ford Island. It is very interesting flying from this historic field. We fly our planes back aboard ship after she is back at sea.

Atsugi, Japan to Guam and Return

On Saturday, 15 February 1958, the day after ATG4 aboard USS *Hornet* (CVA-12) arrives in the port of Yokohama, Japan, our O-in-C, LCDR Bodger, is told that our detachment must disembark immediately, and report aboard Navy Auxiliary Field (NAF) Atsugi for further assignment. Our detachment consists of Bodger plus 7 Lieutenant Junior Grade Officers – 4 pilots and 4 air controllers - and 26 enlisted personnel – including 3 Chief Petty Officers.

After our three aircraft are offloaded from the flight deck, three of us pilots taxi the planes to a nearby runway and fly the planes to Atsugi airfield. Our troops and gear arrive via bus and truck.

As it turns out there are a lot of other detachments and parts of squadrons (split into detachments) of our Air Task Group that are offloaded to fly various classified missions from land stations.

All our flying personnel were fitted with cold-water flying suites while we were at North Island. We had to get into a pool and swim with them on to make sure they did not leak. The boots are not separate, but part of the suite – like booties. We ware these suites when flying over water from Atsugi.

After two weeks at Atsugi, Bodger is told that we will fly to Guam in order to fly classified missions from Agana airfield. Since I am the Detachment Maintenance Officer, all our enlisted personnel work for me. Therefore, Bodger puts me in charge of "my" men and one of our four air controllers to fly on a transport, with all our detachment gear, to Guam. The other pilots and air controllers will fly our three planes to Guam.

Shortly after our arrival on Guam, three North American FJ-4 Fury jet fighters and associated personnel of VF-94 squadron from our Air Group join us on Guam to, ostensibly, provide fighter cover. However, we never fly a coordinated mission.

Our "Classified" mission is to patrol the Mariana Islands, looking for Russian support ships and submarines. The furthest North we fly is Pagan Island, depending upon the amount of investigation we do on the out-bound leg. We carry one 150-gallon drop tank of fuel. We fly irregular missions, including night flights, so our actions are not predictable. We enlarge the scope of our mission to look for unusual shipping, Russian or not.

On one of these missions I turn around before using all the fuel in the drop tank. It is just about time for my engine to cough, indicating it is time to change back to my main tank. So I alert my crew that my engine should cough soon, and not to be alarmed!

About this time we spot a small ship near the coast of an island. So I switch back to the main tank before I dive down to have a close look at the vessel. After we are through with the investigation I climb up to cruise altitude and switch my fuel back to the drop tank. Thinking all of my actions have been standard operating procedure (SOP), I do not say anything more on the intercom.

Soon the engine coughs, and as I am switching back to the main tank, my technician in the back asks nervously – "Are we okay?" or words to that effect.

I apologize to him, saying I thought he would have understood that I would not have dived and flown low above the ocean without switching back to my main tank; and upon returning to altitude, I assumed he would have realized I would have emptied the external tank.

As I recall, this is the only time I ever scared a crewmember enough to have him come up on the intercom!

When it is time for us to return to Japan, I talk Bodger out of my being "in charge" of our men on the way back to Japan. Mitch Mitchell is stuck with being in charge of the transported group.

Since I am the Maintenance Officer I fly all the test hops on our aircraft during the cruise. After all, it is my men who are performing the maintenance work. We complete final engine adjustments and I test fly each of our three aircraft before we head north to Japan. Our planes were named SNAP, CRACKLE and POP before we departed our squadron – each name is painted in small letters above the cowl number.

LCDR Bob Ashford is the VAW-11 O-in-C of Detachment Lima, and they arrive on Guam in time to assist in our departure on Thursday, 29 May 1958. They were also off-loaded from their carrier when it arrived in Japan.

The Navy has a Super Connie (WV2), of squadron VW-1 (typhoon hunters) stationed on Guam, fly ahead of us to a point about halfway to Iwo Jima in order to monitor our flight and guide us, as required. In addition, a destroyer has been positioned in this general halfway area, to rescue us from the ocean, if needed.

It may seem odd, but knowing that someone thinks this destroyer should be positioned there, makes me more, rather than less, apprehensive about this unusually long over-water flight for a single engine, single-pilot aircraft.

NAF Atsugi is forecast to have instrument conditions when we expect to arrive, therefore we must have enough extra fuel to make it to an alternate field plus an hour of holding time. This requires more fuel than we can carry in our main tank and two wing drop tanks, so we must land at Iwo Jima to refuel. Remember, with our radar below the belly of the aircraft we do not have a belly attachment point for a fuel tank.

Iwo Jima, which is the middle of three volcanic islands, is small - about 5 miles long by 800 yards to 2.5 miles wide - is also forecast to require an instrument landing – we use NAF Atsugi as our alternate because there is no landing field near Iwo Jima! And we do have enough fuel to reach NAF Atsugi.

After this island was captured from the Japanese during WW II, it was used to base P-51 fighters to protect the B-29 bombers on there bombing runs over Japan. The bombers (with four engines and pilot, copilot and navigator) were flying from Guam, Tinian and Saipan. Iwo Jima was also used as an emergency landing field for the bombers. But once they landed here they could not takeoff again – runway was too short.

Ed Chybowsky, an air controller, is flying with me on this trip. We have stored our personnel belongings in the rear, where an aviation electronics technician would ride on an AEW/ASW mission.

In the warm-up area, I cycle the fuel selector from "Main" to "Left Aux." to "Right Aux." to ensure all our fuel will be accessible during the flight. Aux is short for auxiliary, meaning a drop tank, with a capacity of either 150 or 300 gallons. The tanks are attached to the left and/or right pylon below each wing. We are carrying two 150-gallon tanks for this flight, because this is all we have. Since 300-gallon tanks interfere too much with our radar coverage, we do not deploy with them.

On Our Way to Iwo Jima from Guam. (By Author)

After our departure, Darrel Westbrook and I join on our leader (Bodger) and fly in lose formation towards Iwo Jima. I switch to my left drop tank as soon as I am in formation, and note the time. There are no fuel gages for the drop tanks, so you must go by the clock and the expected rate of fuel flow. As I learned flying the AD-1 in the advanced training command, I always let my engine cough once before I switched back to the main tank. Then, if two tanks were being carried, as they are now, I switch back to the main for a minute or two before switching to the other tank.

Many pilots prefer switching by the clock rather than the cough – I always felt why not use *all* your fuel rather than carry any extra fuel that you *cannot* use later. When our engine coughs, for lack of fuel, Ed almost panicked – it turns out he is more concerned about losing his personal gear in the back than ditching and being rescued.

Before arriving at Iwo Jima we are flying on instruments. We make individual GCA (Ground Control Approach) approaches to the runway. During a GCA you are following directions from a highly trained operator watching your approach on radar that provides deviations from glide slope and heading to the operator. He vectors you around to get you on the glide slope heading to the runway centerline.

He then keeps telling you how high or low you are in feet from the glide slope (which you cannot see in your cockpit) and whether you are drifting right or left of the centerline. (Neither of these is depicted by our instruments – when ILS, instrument landing system, is available, a new instrument on the instrument panel shows this information to the pilot.) The director keeps up a steady stream of information. What you want to hear, over and over again is: "On heading, on glide slope"

The GCA controller literally talks you down through the clouds. At the height and distance minimums for his field, the controller asks. "Do you have the runway in sight?" If you do not, you MUST execute a missed approach. When asked, I have the runway in site directly in front of the plane's nose, and land without problems.

Because of the low ceiling we cannot see the top of the famous Mount Suribachi – where the five marines raised our flag over this island. Though we do not have time to see much of the island, we do see a lot of aircraft that have been pushed off the cliff at one end of the runway, during and after WW II. Being on this island is a thrill for this WW II buff. I have always been glad that I was able to see and stand on this historic island.

After fuel and oil for our planes and food for us, we make an instrument departure for NAF Atsugi, Japan.

As we near Atsugi field, we are back on instruments and the last 30 minutes we are also flying at night. We each make individual instrument approaches, this time an ADF (automatic direction finder), to a landing at Atsugi.

After flying 8.4 hours I am tired and hungry.

This flight is the most flying I ever did in one day, without a copilot. It is the most memorable cross-ocean trip I ever flew!

Back Aboard Ship

On Friday, 6 June, we fly back aboard ship, which we are told had been designated a CVS (Antisubmarine rather than Attack) the day before. We have no problems getting aboard even though our last carrier landings had been during our brief stay in Hawaiian waters.

I am greatly surprised when I walk on the hangar deck to take care of our Detachment aircraft – I see a couple of VA-216 ADs with some of the Navy markings and US insignias painted over with regular aircraft gray paint! The US insignia is being restored on the side of one aircraft. At the time, I do not get a straight answer for this.

Now (2003) I know that the ship put into Hong Kong on Saturday, 12 April. Then unexpectedly, the word is passed that she is departing forthwith – this is on the 14th. She heads south to the vicinity of Indonesia.

All the ADs have all their Navy markings and US insignia painted over with normal wing and fuselage colors. The aircrews of these planes are briefed on conventional warfare tactics using bombs, rockets and 20mm cannons – the wing guns of the Skyraiders. And they receive a refresher on how to survive in the jungle.

Once in the vicinity, but not within sight of Indonesia, the propeller-driven planes are loaded with conventional weapons, but were never launched. Apparently, they waited about a week for Instructions to launch. On Thursday, 1 May, the ship departs the area for Honk Kong. No information is ever passed to the aircrews as to what this almost "exercise" was all about!

VA(AW)-35 got all their planes repainted prior to returning to Honk Kong. But VA-216 did not – so these planes were hidden on the hangar deck.

For more background information, check on the Internet for US operations in Indonesia during 1958. The CIA has never admitted they took part in any of these operations.

When those of us in Air Task Group 4 (ATG4) departed on this cruise we knew we were not all going to be returning in July – that was just the facts of life in these days. In addition to the 3 airmen we lost during the ORI, we lost 7 more pilots - 3 Lieutenants, 3 Lieutenant Junior Grades and 1 Ensign – and 7 more aircraft from 4 different squadrons – 2 attack squadrons and 2 fighter squadrons.

Awards

USS *Hornet* won the Chief of Naval Operations Annual Aviation Safety Award for fiscal year 1958 for the Pacific Fleet. Air Task Group 4 is commended for the contribution it made in assisting the USS *Hornet* in winning subject award. In turn, each squadron and detachment of ATG4 is recognized for their individual contributions. And finally, I am recognized for my contributions as Maintenance Officer of VAW-11 Detachment KILO.

In addition, USS *Hornet* is named the CVA class winner of the Naval Air Force, Pacific, and Battle Efficiency Competition for the competitive year 1957 – 1958. In turn, ATG4 is commended for their efforts in this endeavor; and VAW-11 Detachment KILO is highly praised for its outstanding performances in the accumulation of a winning score by the ATG4 commander.

Biggest Problem

Absolutely, the biggest problem I found with flying from aircraft carriers is that they were *never* where they said they would be upon my return! I have talked with other old carrier pilots about this and they seem to all agree that this was a big problem. I have read that during WW II it was not uncommon for the planes to return from their mission to the place the carriers claimed they would be, only to find an empty ocean. They would then commence a search to find the carrier – as they were running out of fuel!

During my operational flying (14 January 1957 through 1 July 1959) I flew from five carriers, all were WW II *Essex* class carriers that had been modernized – hurricane bow, angled-deck, mirror landing system and some of them with steam catapults; in addition their armament (guns on and along the flight deck) was completely changed. Also their electronics and other equipment were upgraded. The five were: *Hornet, Hancock, Bonhomme Richard, Kearsarge* and *Shangri La* – in decreasing age. These five were initially commissioned between 29 November 1943 and 15 September 1944.

That was one good thing about flying the AD-5W version of the Douglass Skyraider – we had great radar (APS 20) with which to find the damn ship! Except there were times when our mission required us to remain so low that the range of our radar was greatly reduced, and we had to be relatively close to the carrier to "see" it on radar.

The "best" or "worst" case, depending upon your point of view, of this occurring with me happened off San Diego; at least this is where it started. This was during an exercise from USS *Hornet* (CVA-12) on Wednesday, 9 October 1957, before we deployed aboard her in January 1958.

I am ordered to be part of the morning launch, but am directed to fly from the ship to NAS North Island and then standby for a call to return to the ship. I ask for the ship's intended average speed and direction while I will be gone. I am informed that she will be staying within her present general operating area.

So off we (my two crewmen and I) go for some beach time.

When I check into the VAW-11 ready room, I inform the SDO (Squadron Duty Officer) that a message will be sent from USS *Hornet* telling us when to return. So we have to wait in the ready room, until called. As the afternoon is becoming evening I am wondering what is going on.

The SDO has gotten tired of my asking to double check that no message from the ship has been received at North Island.

Finally, the message comes – "Return to ship." I ask the SDO if the ship has bothered to say where she is – "That is a negative." Which we think means she is still steaming within her operating area. So out we go to the aircraft line, "fire-up" our bird and takeoff for the ship's operating area, which is due west of North Island.

Here I should tell you that there is a band off shore known as the ADIZ (Aircraft Detection and Identification Zone). At this time, you can go outbound from shore, but if you penetrate the zone you cannot turn around and return to shore!

You have to file a flight plan in order to go into the zone and return. It is an automatic grounding if you violate this rule! VAW-11's normal operational area does not extend within this zone.

Likewise, before you enter this zone from the ocean side you must properly identify yourself and have filed a flight plan.

Upon turning outbound towards the ship's operating area, westward, I tune in the low frequency radio compass to the ship's frequency. The needle immediately points to the North-Northwest, which is *not* in the direction of the ship's operating area. I did not know if my radio compass is malfunctioning - the frequency I dialed in is not the one being used, the direction of the needle is erroneous or what?

So we keep heading west to the ship's operating area. I have climbed to about 8,000 feet to allow our radar to have a good view around us. My AC (Aircraft Controller) reports he has nothing positive in the way of our ship.

Based upon my DR (dead reckoning) navigation, we are now getting close to penetrating the ADIZ – outbound. So I finally tell the crew I am going to assume the radio compass is working and that we are going to follow it to see where it leads us!

We turn almost due north to fly the needle towards the source.

We are now definitely outside of the ship's operating area – based upon what we had been told in the morning about the operating area – and going further away from it! I conclude we have penetrated the ADIZ, so there is no going back.

Shortly the AC reports he has what might be the ship dead ahead of us. With a feeling of relief I keep going towards the source of the radio signal.

Finally, we see a carrier – but, is it ours?

I call the ship on our ultra high frequency (UHF) radio (which has a line-of-site range, while the radio compass may have a range up to 200 miles), using our respective call signs. Yes, it is our ship; they have us on radar, and tell me to make a straight-in approach to a landing.

I had never heard of a "straight-in" being used aboard ship with propeller-driven aircraft. What the *HELL?* The jets would use this approach to conserve fuel.

A normal landing circuit for us is to fly low down the starboard side of the ship (at about 500 feet) on the ship's heading (check your compass heading), break left when a short distance in front of the bow. Keep it fairly tight in a 180 degree descending turn, rollout out on the reciprocal heading of the ship and start performing the last of the landing check list by getting completely "dirty" – hook down, wheels down and flaps down – before commencing the approach from a point known as the "180" – because the pilot must fly through 180 degrees before landing – that is a point 90 degrees off the port stern of the ship. Just before starting the approach, you identify yourself with your call sign and report: "Flaps, wheels, hook, down" (We never reported the fuel onboard unless we were very low on fuel, or requested to do so because PRIFLY thought we were getting very low.)

I fly what I think is proper for a straight-in approach and get aboard on the first pass.

As it turned out, this is my first and last straight-in approach to a carrier in a propeller-driven plane. I think they gave me a straight-in because the ship is in a hurry to secure from flight operations; ours is the last landing of the day.

In the ready room I am asked, "What took you so long?"

No, I do not know why they ordered us off, or why we were gone for so long, or why the ship was so *far* away (over 100 miles) from where I had been told would be her position!

Such is the life of a junior aviator aboard an aircraft carrier.

10
VAW-11 – Post-Deployment

Non-Flying Duties

Since I had been Maintenance Officer on the WESTPAC Cruise, I had assumed that I would be assigned to the Maintenance Department upon my return to the squadron; so I am surprised when I am assigned to be the Ground Training Officer in the Operations Department, still headed by a Commander. Knowing what a pain it was to see missing pins on the Training Board when almost all the training has been completed; I propose to the Operations Officer to reverse the scheme of "pinning." Put all the pins in at the beginning, and remove them as the training is accomplished!   Believe it or not, I have a hard time to convince him that my scheme will make our jobs easier. The Navy is known to be very slow to make changes – this is no exception! Finally, he relents and lets me try it – but, if he does not like it, we will go back to the "old standard" way! Not sure if part of the problem is that I'm a LTJG and he is a CDR.

As Ground Training Officer I am responsible for scheduling all the required classroom training of pilots, air controllers and aviation electronics personnel. Double-checking that it has occurred and who all attended, keeping all the training records up to date, and pulling out the pins on the Training Board, I have no assistant.

Several months later, the Operations Officer congratulates me and says the new scheme is much better than the old one!

Shortly after returning from cruise I receive a letter from the Bureau of Naval Personnel informing me of my selection for retention in the regular navy. This is what I anticipated, but it still makes me happy to know I'll be flying many more years.

Jackie delivers our second child in a civilian hospital in Coronado, California during the latter part of August 1958. We're sure glad my cruise was not extended!

During February 1959, Darrel and I are tapped, at separate times, to attend the Military Justice course taught by the Fleet Training Center at the Naval Station in San Diego. The one-week course mainly explains the Uniform Code of Military Justice (UCMJ). The UCMJ is relative new to all the services. Later we are assigned additional duties – Defense Council for me and Trial Council for Darrel. Occasionally we oppose each other in Special Courts Martial.

The lowest Courts Martial is Summary, only one officer performs all the functions of defense, trial and judge – this is for serious, but minor offenses. I never had to do this. The highest, General Courts Martial, is for very serious crimes; which requires lawyers, for Judge and both Councils. However, the Judge only rules on points of law and does not conduct the trial. The senior officer of the Board (jury to you) conducts the trail in Special and General Courts Martial.

The lowest level of punishment, which is only for enlisted personnel, is non-judicial, called a Captain's Mast in the Navy.

On one occasion when I am Defense Council, a Lieutenant Commander (LCDR) in the Operations Department is the senior member of the Special Courts Martial Board. I am able to have the Board find the defendant not guilty of stealing the five "stolen" items found in his locker during a surprise locker inspection in the enlisted men's barracks.

After the trial the LCDR asks me if the defendant is really innocent. But his question to me is worded in such a way as to convey to me that I should not have gotten him off if he were guilty! This shocks me that a LCDR actually thinks a Defense Council should determine if a person is guilty or innocent, and defend accordingly – now I knew why the Navy and the other services needed UCMJ – so all defendants will be treated fairly and equally in the various services!

Frankly, I thought my client would be found guilty. How obvious could it be that these items did not belong in his locker, including items belonging to his shipmates and the government? But my job is to defend him, not to judge him! I told the LCDR that I did not know; which is technically true. The LCDR seemed surprised, but let it go at that.

Bad News

Normally new aviators spend their first four years of operational flying in two different squadrons. However, after returning from our cruise aboard USS *Hornet*, we learn first time aviators sent to VAW-11 spend *all* four years in this one squadron and the only way to depart early, with a good fitness report, is to attend the Naval Postgraduate School (NPS), in Monterey, California. Since the operational flying in this squadron is not as challenging and thrilling as in an attack squadron, I apply to attend the NPS, thinking I will receive my Masters Degree in Aeronautical Engineering. However, I list other courses on my preference sheet that look interesting *and* offer a MS degree. I happen, at the last minute, to add Meteorology on the bottom line for preferences.

Pick Up Overhauled Plane

This is not a post-deployment story, but should precede the following story, which is a post-deployment tale.

Another junior pilot duty is to fly north to NAS Alameda in the San Francisco bay area, to pick up an AD –5W or –5Q from the Navy Overhaul Facility located on this base. Another pilot and maintenance Chief goes with you. At Alameda, the Chief checks all the paper work and once he is satisfied, the pilot who is going to fly the plane back to NAS North Island performs a preflight check and then flies a prescribed test flight. Of course, if the plane fails its test flight it stays there until it does.

On my first pick-up flight the second pilot is senior to me, so he gets his choice as to when he is going to do the flying. For reasons that I do not remember he chose not to do the flying to Alameda. Since both pilots always fly in the front seats, the Chief rides in the back where the electronic technician normally rides.

So I am going to fly our plane to Alameda, test hop the overhauled plane and fly it back to North Island. I file an IFR (Instrument Flight Rules) flight plan because the northern part of California is cloudy and expected to remain so. Sure enough, I am on instruments before we make our left turn to proceed to the Oakland area.

In the Oakland airspace I am told by the FAA controller to hold at "XYZ" radio beacon at my present altitude, because of traffic at the Oakland Airport, with a specified time of departure. The time of departure is not his best estimate of how long we will be holding, but is a modification of our original clearance given us before we took off at North Island. In case of radio failure we are now cleared to commence our approach at this specified time.

I enter the racetrack holding pattern in accordance with the specifications of the published Alameda approach plate. Ours is the only plane holding at this time. There is minor chitchat on the intercom between us pilots while we have gone around the pattern several times.

All of a sudden, the Chief speaks up on the intercom and says we are cleared to depart the holding pattern and commence our approach to Alameda. I ask him how he knows this. He replies he heard it on our radio. After some more discussions, we (both pilots) individually test our radio connection (we only had one UHF radio in the plane) from the front and find we are both "dead" with respect to transmission and receiving.

I tell the Chief to inform the FAA air controller about our radio predicament and that we will commence our approach in "X" minutes. The "X" minutes is the time it will take us to reach the beacon heading inbound. I tell him to repeat on the intercom *exactly* what he is being told on our radio.

So, based upon the word of our Chief, I return to the beacon and commence an instrument approach to NAS Alameda. When the Chief informs us that he has been told to contact the Alameda tower, the other pilot switches our radio to the Alameda tower frequency. I instruct the Chief to contact the tower and inform them of our radio situation and request permission to land. The Chief soon tells us that we have been cleared to land.

We brake out of the clouds a few miles from the field, aligned with the duty runway. As a back-up to the radio transmissions, the tower personnel are nice enough to give us a green light – meaning we are "cleared to land" – with the tower's hand-held spot light or lamp.

Next morning the weather has cleared and I prepare to fly the test hop on the overhauled plane. The Alameda Duty Officer briefs me as where the test flight is to be flown; which is north of the Golden Gate Bridge over the ocean.

It is fun to fly from NAS Alameda and see all the bridges up close and personal. Once I am in the test-flight area I fly the required tests, as specified by VAW-11. And I get to see the bridges of the Bay area again upon my return to the field. Both planes are "Up" (our radio problem has been fixed) and we have no more problems on our way home –

flying a loose formation – to North Island. I am flying alone, because the Chief does not want to fly in the overhauled plane.

I learn later that this is the response of *all* our maintenance Chiefs who make these flights. They do not want to fly in a plane that has not been carefully checked by the Maintenance Department personnel of VAW-11.

Later, as Maintenance Officer of VAW-11 Detachment Kilo, I hear all sorts of horror stories about what all are wrong with the planes coming out of the Navy Overhaul Facility at Alameda. This is the only facility on the west coast that overhauls ADs. However, I never hear of any problems from the pilots who flew the planes back to our squadron!

To The Bone Yard

With this background I think you can appreciate that the most apprehensive flight I have to take, while in VAW-11, is shortly before I leave this command. I have to fly one of our planes to the "bone yard" at Litchfield Park, Arizona. I feel that our Maintenance Department has swapped every good part for a bad one on this plane and has installed the highest time engine in the squadron on this plane before it is destined for its last flight – at least for the Navy.

These flights are only made in the daytime and in good weather; which should tell you something about the condition of the planes. The only good thing about the flight is that I am being escorted. At least a fellow pilot will know if and where I go down! We return to North Island in this escort squadron plane. Luckily, I have no problems, but some of our other bone yard pilots did.

Pull a Trick

As you may or may not have gathered, except for our carrier flying, the flying associated with VAW-11 is rather routine for young pilots trained for attack flying!

Well, at the conclusion of one practice AEW (Airborne Early Warning) flight for our air controllers and electronics technicians, Darrel Westbrook and my planes end up being close to one another when it is time to head for home from the vicinity south of Santa Catalina Island. Darrel joins on me at an altitude of about 5,000 feet as I head for North Island.

When the leader's radar is active, we fly formation in a stepped-up manner. That is, the wingman is flying higher than the leader. It is suspected that the radar beam at formation range is not good for your health. Usually the wingman fly's to the right of the leader because the pilot sits on the left side of the AD-5W. Since our radar has been turned off, Darrel is flying stepped-down, and for some reason had chosen to be on my left.

Brief examples of the power of our radar beam follow: Once, as Maintenance Officer of Detachment KILO when I am watching the repair of one of our aircraft at a bay on the hangar deck of an aircraft carrier, my people give me a demonstration. One of my electronics technicians picks up a fluoresce tube and places it within the beam of the radar, which is pointed out over the water from the bay. It lights up as though it were connected to an electric circuit. On the carrier we have to be very careful with the radar beam. The radar is never turned on (meaning the radar is emitting a beam) on the flight deck or when we are near the ship. There is real danger of the beam firing rockets, etc. from the wings of the planes on the deck. I heard that on one carrier a rocket was fired in this manner; fortunately no real harm was done because the rocket didn't hit another plane or the ship.

Sorry for the digression.

As Darrel and I are flying in a southerly direction back to base I see a large ship near the horizon that is moving in an easterly direction. As I have said in another story, the wingman flies on the leader and often does not know what is happening "in the real world." So, I very gently ease the nose of my plane slightly down and slowly adjusted my heading for an initial point slightly in front of the ship so we will pass directly over it.

I fly very steady so I will not raise any suspicions on Darrel's part.

As we are approaching the surface of the water I again very gently level off just high enough for Darrel to safely clear its superstructure in level flight – but not with a lot of extra space. But this will be no fun, for he will never see the ship until we are passing over it, if at all.

So, just prior to the ship I pull up sharply, knowing he will stop flying formation and look straight ahead to see what the reason is for such a violent maneuver.

Since Darrel does not call me on the radio, we are still on one of the tactical frequencies assigned to VAW-11; I do not know how he feels about my tricking him.

We are nearly back to San Diego, so Darrel does not rejoin on me for the final few miles. ADs are not allowed to fly formation around the field, unless they are going to land on the runway. As I have stated, in good weather the air station prefers we land on the mat and not the runway during the day.

When we are back in the VAW-11 ready room we are both laughing about the incident. He tells me that he thought we were still at 5,000 feet and was startled to see a big ship right in front of him!

He is not upset, because he knows I'd never place him in danger! However, he is surprised that I could have taken him from about 5,000 feet down to about 100 feet without his knowledge, even if he were flying formation. He thought I had done an excellent job of flying to accomplish this. He appreciated my joke!

Vertigo

Unfortunately, in the AD-5W the magnetic compass is located at the top-center of the windscreen. In this location it is completely useless during instrument flying because it is *not* within the normal instrument scan when flying in IFR conditions. One flies a heading based upon the gyrocompass, but if this instrument fails, it would have been nice to have the standby magnetic compass somewhere on the instrument panel!

Likewise, for some unknown reason, the AD-5W had a second gyro horizon in the bottom-center of the instrument panel pedestal between the pilot and aircraft controller. The bottom of this instrument's dial is at floor level. In this location it is completely useless, absolutely outside of the instrument scan.

When flying on instruments, one must move the head as little as possible to prevent vertigo. You scan your flight instruments only with your eyes. During vertigo your body (mainly your ears) is telling you that you are not flying straight and level when you are; or that you are not banked and in a turn, when you are.

Fortunately, I never experienced vertigo while flying. I have experienced it while in a flight simulator and made head movements that you are not suppose to do when flying on instruments – except I was, at this time, in order to experience vertigo and understand it can happen.

One time, when Darrel and I fly from North Island to NAS Alameda to pickup an overhauled AD-5W, I am doing the flying and he is in the right seat. We have a Chief mechanic in the rear seat. I am flying on instruments in the clouds, and when I make the left turn to change airways and head towards Oakland, Darrel has a strong attack of vertigo – which takes him a long time to shake. Out of the corner of my right eye I can see he is really fighting it! Fortunately he trusts me enough not to make a grab for the controls, etc. I figure we are now even for my taking a cat shot with him doing the flying.

Accepted by NPS

I am very happy upon learning of my acceptance to the NPS, but it quickly fades to disappointment upon discovering I will be studying Meteorology rather than Aeronautical Engineering. I go to the weather office on base to get good tips on the Meteorology course of the Postgraduate School.

Surprisingly, rather than congratulations, the Lieutenant informs me it is a big mistake for an operational aviator to study meteorology at NPS because of the payback tours required for graduating with a degree in meteorology. This is the first I have heard of any payback tours, all I have read is that the Navy needs and wants officers with advanced degrees.

The Lieutenant Commander returns to the office and after introductions, he says the same as the Lieutenant. Finally the Commander, in charge of the office, returns and his comments are identical to the first statements - payback tours are a career killer for the operational aviator. All three are aviators.

They *each* recommend that upon arriving back at my squadron I should immediately request the Bureau of Naval Personnel to rescind my orders to study Meteorology. During the next few days I decide it will do my career more harm than good to try and get my orders cancelled, so their unanimous recommendation is not taken.

Engine Trouble

During a standard approach to a landing, if your engine quits anywhere except just before landing, you will *not* make the runway or flight deck! When you practice a deadstick landing (meaning your engine is not running – don't ask me why it is called "deadstick"), you fly at least twice as high as you would with a normal landing approach.

In the AD-1, we would fly about three times as high and while approaching the runway cut the engine (pull the throttle back to idle), then pop the dive-brake under the fuselage and dive for the runway, retract the dive-brake, flare and land three-point on the runway.

In the AD-5W, which does not have any dive brakes, you cut your engine wherever you want to start your deadstick landing. Your height has to be high enough to be able to glide to a landing. Because of the unusual height of the approach you have to get permission from the tower to practice this "emergency" landing; which we had to perform periodically.

While flying south over San Diego on a simulated instrument approach for a landing on runway 18 of NAS North Island, my engine quit, then came back alive, only to quit again before I get it running again.

During the landing checklist one moves the Mixture Control forward from "Normal (Auto-Lean)" to "Rich." However, every time I moved it to "Rich" my engine would quit!

One thing that you learn early in flying is: *If something bad happens after you have done something, especially if it happens twice, you don't do that anymore – regardless what the flight manual says!*

So I call the tower and tell them that I am having some engine problems and that I will be making a precautionary higher than normal approach to runway 18.

Just off the end of the runway is a damn ship moored to the pier! Sure wished it hadn't been there! This means I must be high enough to clear its masts, if I am forced to glide to the runway.

I warn my crew to prepare for a crash landing, but I think we can make it!

Once we are over the ship, I begin easing off the power and make a smooth landing; only it is much further down the runway than normal.

On the Yellow Sheet, I describe my engine problem and "Down" the plane.

The next day the Maintenance Officer asks me if I had performed the "Idle Mixture Check" during the "Takeoff Checklist?" I report I had (because I *always* do all the checks, before flight). During this check, the Mixture Control is moved slowly from "Normal" towards "Idle-Cutoff." The RPM should rise slightly before the engine starts to quit. No rise indicates the idle mixture is set too lean, and a rise of over 10 RPM indicates it is set too rich. In either case, you do not fly the aircraft.

He says the mixture is way too rich for the engine to run in "Rich" and is surprised I had not had problems on the takeoff – you takeoff with the mixture in "Rich." In fact the mixture was so rich; he was surprised the engine was even running in "Normal!"

He did not understand what had happened to the engine during flight to get the mixture setting so far off!

Scare Your Flight Leader

LCDR Bob Ashford and I have flown two AD-5Ws, with one passenger each, from North Island to Monterey to check on availability of housing while we attend the Naval Postgraduate School (NPS) during our next set of orders. I don't remember why the passengers have come to Monterey – we do not see them again until it is time to leave, the next morning.

After I "fire-up" my engine to return to North Island, the gyro horizon fails to start working. I "CAGE" it and "UNCAGE" it, and it still fails to work. Since Bob has filed a VFR (visual flight rules) flight plan it is legal for me to depart without this instrument working. That is, in VFR conditions one is not required to have a gyro horizon, because one has the natural horizon.

So off we go to NAS North Island.

As we approach San Diego we see clouds ahead. Soon we are told that North Island is IFR (instrument flight rules). We are switched from enroute to approach control for approach instructions. I am tucked in tight on Bob's right side. There is no way for me to inform Bob that my gyro horizon is malfunctioning. The AD-5W has only one UHF radio, and pilot-to-pilot talk is *not* allowed on FAA frequencies.

As we get closer to entering the clouds I realize this is *not* a good idea for me to fly formation in the clouds, with the possibility that if I lost sight of him I would be forced to make a rapid transition to flying on instruments, without my gyro horizon. In addition, Bob has agreed to a section landing on the duty runway.

During a section landing, when in right echelon, which we are in, this requires the lead aircraft to take the left side of the runway and to hold off on his actual touch down. The second aircraft takes the right side of the runway and tries to get on the runway as soon as possible. Of course the *main* consideration in all of this is that it should be briefed *before* the flight, so there is no misunderstanding as to what is going to happen. We had never briefed it, and I had never practiced this with anyone before – I just knew what should happen, but did not know what Bob is planning to do.

In addition, I am not sure if it is legal for me to go from VFR to IFR in civilian airspace without my gyro horizon. Had it malfunctioned when we were already in IFR conditions, then it would have been legal for me to continue.

Therefore, I break off from Bob just as we are entering the clouds, with a climbing right turn. Bob goes on his way and his passenger never tells him we are gone – which a good passenger, that is properly briefed, would have done. Bob executes the required instrument landing approach and as he breaks out of the clouds the North Island tower, expecting two aircraft, tells him they have only one aircraft in sight.

Bob thinks I must have gone in (crashed) somewhere and is really concerned for me and of course my passenger, who is a flight surgeon stationed at North Island.

Meanwhile, back in my plane, I fly VFR above the clouds to North Island and call their GCA (Ground Control Approach) Unit. I might add that in military airspace, which I am now in, it is legal for me to go from VFR to IFR under GCA control without my gyro horizon.

I tell the controller my gyro horizon is out and request a GCA approach into North Island with level turns. This of course, I have practiced many times – without the level turns and with all my instruments working. However, I know I can do what is called "partial panel" (no gyro horizon – after all, in the real old days of flying there was no gyro horizon to be installed).

I am going to use my turn-and-bank indicator, altimeter, rate-of-climb indicator and compass to fly the pattern requested by the controller. All my turns are to be level, to simplify my instrument scan. I had never flown partial panel in an AD, but I had in SNJs and T-28s, with an instructor in the front seat.

I probably should not say this, but I am relishing the fact that the flight surgeon – whom I did not know – is really scared! Now he knows how we pilots feel during our annual flight physicals!

There are no problems during the approach and landing, though it is a low ceiling, and I am glad I have not tried to fly in on Bob's wing.

In the VAW-11 operations building, Bob reports to the SDO (Squadron Duty Officer) that we might have crashed, so he is really glad to see us, when we walk into the squadron spaces.

Tragic Error

Shortly before or after I returned from deployment on the *Hornet*, about 2 July 1958, our squadron gets a small group of jets. They are the F2H-2 Banshees, which first flew in January 1947 and are stricken from the Navy in September of 1959.

VAW-11 had been VC-11, a composite squadron – which means significantly different types of operational airplanes are in the same squadron. Someone got the idea that the Banshees should be assigned to ASW (antisubmarine warfare) carriers (CVS) to provide fighter protection for the ship and ASW planes (S2Fs). However, why VAW-11 was selected to have the fighters is unknown. Some of the detachments of fighters did deploy, after I had departed the squadron.

The AD pilots were informed that they would not be flying the F2H!

If I remember correctly, the senior jet pilot is a senior Lieutenant, and is the O-in-C of the first jet detachment. I do not know how many enlisted jet qualified maintenance men the squadron also gains. VAW-11 has about 100 planes and well over 200 officers – it is not the typical Navy squadron! Our counter part, VAW-12, is at Norfolk for the Atlantic Fleet.

One day the senior jet pilot and a wingman go on a flight to practice instrument work.

Later in the flight, the leader decides he and his wingman will execute a jet penetration and landing approach to a field north of San Diego. There is a thick layer of clouds below them so the practice penetration will put them on instruments. The wingman is depending upon his leader to keep him out of trouble.

When cleared by the FAA they commence their high altitude penetration in accordance with the jet approach plate – listing headings, altitudes, turns, etc. Late in the penetration the wingman sees something dark in his peripheral vision and looks ahead – trees!

He pulls as hard as he can on his stick to pull out of the dive – jet instrument penetration. If I remember correctly, his plane flies through the tops of a few trees, but does not crash – the same cannot be said of his leader!

This is the first funeral Jackie and I have attended since our marriage. He is buried on the Point Loma Peninsula. The experience is very sobering – taps sounded so *mournful!*

The F2H accident investigation revealed the gyro of the gyrocompass had been installed improperly in the wing, resulting in a significant error – 20 to 30 degrees (as I recall) – in the gyrocompass. This compass error was enough for him to miss the valley in the ridgeline of the mountains on the approach.

One of the first checks that a pilot should perform when he is positioned for takeoff from a runway is that the magnetic compass agrees with the runway magnetic heading (within a few degrees) and the gyrocompass agrees with the runway true heading (within a few degrees) or corrected for variation agrees (within a few degrees) with the magnetic runway heading. Evidently the senior jet pilot omitted the compass checks after he pulled into position for takeoff on the runway. Even though the yellow sheets for his plane should have listed this work having been performed on the gyrocompass, which should have made him doubly alert to possible problems with this instrument.

His wingman apparently was tucked in too tight to double-check his instruments – that is one problem in tight formation flying.

Put To the Test

After my WESTPAC cruise, my flying duties include assisting new pilots transitioning from flying the single seat attack version of the Douglas Skyraider to our version, the AD-5W.

Ensign Buckner has completed all but one of his daytime FAM (familiarization) flights, with the assistance of other squadron pilots. These flights have primarily involved shooting touch-and-go landings at Ream Field. On Thursday, 16 April 1959, I am riding where the air controller normally sits, the right seat, with Buckner as pilot. We complete his area orientation flight (which is normally the first or last FAM flight) and head back to base.

His flying is too tentative for my liking. So I tell him to simulate an engine failure (by pulling the throttle back to idle) and inform me why he is performing each task and which options he is considering. After we are back to normal cruise, I explain how he could improve his procedures. As a check to see if he has fully comprehended my reasoning, I ask, "What would you do if your engine quit in the 'break' for a landing?"

During visual flight rules, carrier type aircraft fly down (into the wind) the duty runway with excessive speed and break (a sharp left turn) somewhere between 3/4th and the end of the runway (traffic permitting) - this is the up-wind turn to the down-wind leg of the landing pattern. The break at a Navy field is always to the left, because this is the direction of a carrier break.

Buckner is making the same mistakes he made before by being indecisive about the sequence of events he would follow and what he would consider in selecting an emergency landing site. When he says he would ask the control tower operator for permission to land off the duty runway, I cut him off with a stern remark, "You are *not* in the training command anymore!"

Then remind him that when he has an emergency, his first and foremost responsibility is to *fly* his airplane! Never ask permission for anything. Do not be indecisive! Quickly run through your options - which hopefully you have thought about before - pick the best one, and go for it! Tell the appropriate person what you are doing.

In addition, I explain runway 27 is in use most of the time, especially this time of the year. Since Point Loma Peninsula (which runs north-south) is off the end of runway 27 there are two choices for engine problems in the break: ditch in the ocean to the south, or continue your turn and land on the large asphalt area of the field. If you choose the asphalt, do not lower the landing gear unless you know the field is made with height to spare *and* there is still time to get the gear fully extended.

The question about an engine failure in the break was chosen for two reasons: 1) this is about the worst place to have an engine failure if you are indecisive, you are low and do not have much time to make a decision but normally have several choices, and 2) whenever the throttle is significantly, rapidly changed the probability of an engine malfunction is increased.

The next day, Friday, a sailor who is new to the squadron (and I think is just out of boot camp), comes up to me at quitting time asking for a right seat passenger cockpit check-out. I tell him no, that it is too late and I will do it Monday. But he pleads with me, that he had gotten permission to fly, has located a pilot who will take him up over the weekend, and he really wants to fly!

He was informed that I, as Ground Training Officer, had to sign off his chit indicating that he had had the checkout before he could go flying. He hinted that I could sign it off without taking the time to actually perform the checkout.

I finally relent, and take him over to a plane in the hangar and go through the right seat passenger cockpit checkout. I go slowly and deliberately because I can tell he is very excited about getting to fly and I want to make sure he understands each of the instructions. This instruction and hands-on demonstration includes emergency procedures. I instruct him where and how to connect his oxygen mask and how one tells whether it is working properly, etc. However, he does not have a mask with him at the time. So I tell him to be sure and have the pilot demonstrate this to him before he flies. I tell him at least three times to be sure and have this done *before* he flies. Each time he promises he will.

Upon arriving at work Monday morning, I hear that we lost a plane Friday night on a local flight. Later Lieutenant Wheeler tells me that he was the section leader of Buckner's first night FAM flight. They had completed the flight and were in right echelon for the break on runway 27. He had made his break and soon he heard Buckner on the radio telling the control tower operator his engine had blown and he was going to land on the asphalt mat. He said Buckner had crash-landed on the field and is in the Balboa Navy hospital with burns.

Later I find out that Buckner had a crewman with him on the flight, and that he was dead. This surprises me, because I thought it is squadron policy that passengers are not allowed on FAM flights. Later still, I find out that the "crewman" was the sailor who had been so eager to fly.

During the week I find out that when the crews of the crash trucks arrived at the plane, Buckner was out of the plane and his passenger was still strapped in his seat in the burning plane. They suppressed the flames in the cockpit area and someone checked the passenger and reported, "He's dead." They kept the fire out of the cockpit and left him strapped in the seat of the plane as they extinguished the fire and attended to Buckner. The crash landing was next to North Island's munitions bunkers. Officially the crewman had died of asphyxiation.

Saturday, while visiting Buckner in the hospital, he says the engine blew up when he throttled back in the break. He did not want to attempt ditching at night, so he elected to try for the field. The fire came through the firewall and burned his lower legs and wrists during the approach. He did not lower his gear until he felt a wheels-down landing could be made. When the plane came to a stop he pulled the quick release (of seatbelt and shoulder harness) - and fell on his head!

He had not realized the plane has flipped upside-down. Apparently this was caused when the main landing gear clipped the top of the chain link fence along the outside edge of runway 27.

I did not ask him any questions about his passenger and he did not volunteer any information.

I am proud of Buckner; he made a good decision – without hesitation - told the control tower operator what he was going to do, and saved his life!

But I have occasionally wondered if they would have both been dead if I had not followed my instincts and probed his procedures for handling an engine emergency.

Flying My Father

The Navy calls the twin engine Beachcraft, which carries two pilots and up to six passengers, the SNB (Super Navy Bomber – its joke name); the Air Force calls it the C45. It is a utility aircraft for both services. It is smaller, in terms of wingspan, length and height, than the AD-5; and has only 900 hp, 450 hp each engine, while the AD-5 has 2700 hp in its one engine.

Beachcraft SNB (US Navy)

After learning that I am going to the Naval Postgraduate School (NPS), I hear that the Monterey Navy Auxiliary Field has only SNBs for me to fly. Therefore, I request to become qualified in the VAW-11's SNB-5. It doesn't take much to qualify in the SNB. But before I am qualified as First Pilot in the SNB, my parents visit us. I make arrangements with my Commanding Officer for Dad (who is still performing Air Force Reserve active duty during the summers) to fly in the SNB, after he signs a waiver in case of an accident.

So "Mitch" Mitchell, who is qualified as first pilot (he was our fourth pilot in Detachment Kilo), and I take him flying. I let him sit in my copilot seat after we are airborne. He really enjoys the flight, but is rather subdued after we land. However, when we get home and he begins telling Mother about the flight, I can tell from his story that he is really thrilled he has flown with me.

I am disappointed that I could not have taken him up in an AD, as long as he had to sign a waiver anyway. However, I never knew of any of our ADs carrying a civilian passenger.

Author as a LTJG in flight jacket with VAW-11 squadron patch.

11

NAVAL POSTGRADUATE SCHOOL

MS Student

While checking in at the Postgraduate School during July of 1959, my promotion to Lieutenant is confirmed by the Personnel Department. Upon reporting to the Curricular Officer of Meteorology and Oceanography, Captain Ted Harding, I learn everyone has their academic record evaluated to determine if prerequisite courses, mainly math, have been taken and if the grade point averages, in those courses and overall, are high enough to qualify them for entering the program for earning a Master of Science (MS) Degree in Meteorology, with a strong minor in Oceanography. In addition, each student is interviewed. Commander Leo Clark, Assistant Curricular Officer, interviews me.

Roughly 45% of the meteorology students are permitted to enter the MS program, the others enter the Bachelors of Science (BS) program. Another 20% in our initial MS students drop back to the BS program during the first year. This education program is open to foreign naval officers.

Later we learn that anyone who requested Meteorology on their preference card, first, last or wherever, was selected for Meteorology – the navy is short of meteorologists!

Our third child is born in early August 1960, in the nearby Fort Ord Army hospital.

The first computer we learn to program, in machine language, is the National Cash Register model 102A. We generate the computer instructions on punched paper tape. For my thesis research I am using the leading scientific computer of its time – the Control Data Corporation (CDC) 1604; which is occupying a whole room. We are still using a paper tape reader to load the program, which is written in assembly language, and we must learn to operate the computer. The school is not furnishing computer operators. This computer is installed at the NPS, but it must be shared with the navy personnel working on the development of the first navy atmospheric forecast numerical model – the start of Fleet Numerical Weather Central – a name that it will have when it becomes operational.

Unfortunately, two or three of our foreign students do not earn their Masters Degree. I miss being number one in our class by a few thousands of a grade point. However, we are both nominated and accept membership in the honorary Scientific Research Society, Sigma Xi.

My MS thesis, "Minimum-time Ship Routing Using Calculus of Variations," is published. It is the leadoff article of the first issue of a new American Meteorology Society publication – "Applied Meteorology." This is the first time that calculus of variations has been used to solve the ship routing problem. I have to use simulated ocean wave conditions, which change enroute, because no ocean wave model exists!

Shortly before graduation I learn my first payback tour for studying meteorology will, unfortunately, be directly after graduation, rather than after another squadron tour. I am *very* disappointed that I will not be returning to a squadron.

Flying

Since I qualified as first pilot in the SNB while in VAW-11, at NPS all I need is a check ride on Saturday, 22 August 1959, to ensure that I am safe to fly their SNB-5s and to learn the local area and protocols at the Monterey Airport before being scheduled to fly the SNB. The civilians began sharing this field with the Navy during WW II. In addition to logging flight time, including required instrument and nighttime; we fly naval personnel, either outbound or inbound, to various military fields in California. One day I am flying as first pilot with a fellow classmate, Bill Stevens, as copilot to NAS Point Mugu to pick up some NPS students who have flown down earlier and are waiting for a ride back to Monterey. We are flying VFR (visual flight rules) at a lower altitude than will be required to fly over the mountain range at the end of the central valley of California to reach Point Mugu. I wait until we are almost at the end of the valley before I start climbing. I add power to climb, notice we are climbing by the rate-of-climb and altimeter instruments. However, soon

I see we are losing altitude! So I add more power to get us climbing again. Soon we are losing altitude again. I go to normal rated power (maximum power – except for takeoff) and start climbing again, but soon we are losing altitude again! Naturally I have heard of the downdraft one can experience around mountain ridges; but this is my first experience up close and personal. I am really amazed that the SNB cannot climb at full power with just two crewmembers onboard! Of course this is not dangerous; I just do a 180, and soon start climbing again. But it is an eye opener for me, flying around mountains with a low powered aircraft.

The Navy SNB-5 is a taildragger, but the Navy does not land it 3-point. Supposedly the tail is not strong enough for this type of landing. So we land it the same way the Air Force lands all their taildraggers – on the main wheels. In the SNB, it is a little tricky to do without bouncing. I've forgotten the exact details, but as I recall one lands with the tail *slightly* down and immediately upon touchdown forcefully but incrementally starts pushing forward on the yoke – yes it has a yoke rather than a stick – to hold the plane on the runway with the tail up. If the plane is level or slightly nose down upon landing, you usually bounce back into the air – which is not fun and the recovery can be dangerous if not handled properly and swiftly.

As in any taildragger, upon landing one *never* brakes with *both* main wheels at the same time! The reason I mention this is that a fellow classmate (in the BS Meteorology program) ends up putting his SNB on its nose at the Monterey Airport. Yes, the plane stays in this position. I see it shortly after it happens, while making a landing on another runway, but never hear the reason – he is not forthcoming. He was landing on the shortest runway; which has since been removed.

We both go to Guam from NPS, I report to FWC/JTWC Guam and he checks into squadron VW-1 - which flies the navalized version of the Super Constellation, WV2 (EC121K – later designation). He is never able to qualify as plane commander in VW-1. An Air Force weather squadron and VW-1 jointly fly the tropical depression, storm and typhoon location and penetration missions in the Western Pacific.

The insulting part of flying while being a student at the NPS is that the Navy aviation detachment at the

Link Trainer (By author)

airport periodically makes all aviators "fly" the damn Link Trainer! What a waste of time! I have not "flown" this trainer since leaving Corry Field in the Basic Flight Training Command.

12
FWC/JTWC Guam

Having had to select a meteorology billet for my next assignment, I at least get my first preference choice, Fleet Weather Central / Joint Typhoon Warning Center (FWC/JTWC) Guam, in June 1961. We fly in a Lockheed Super Constellation from Travis Air Force base outside Sacramento, California to Oahu, Hawaii, to Wake Island and finally to Guam. We have short rest stops on Oahu and Wake Island.

FWC/JTWC is in two large Quonset huts near the COMNAVMAR headquarters building, a few miles SW of Agana, on top of Nimitz Hill.

After standing a couple of familiarization watches, I become a FWC Forecast Duty Officer (FDO) for all of the Western Pacific Ocean! Fortunately, we mainly forecast for the Northern Hemisphere. My assistant is a senior Aerographer's Mate First Class (AG1) Ableman; the other three FDOs have Chiefs. All our preparation work is done manually. My watch section includes communications personnel, who gather incoming weather data messages for our plotters, transmit our analyses of this weather data by fax and send out our forecasts as messages. Most of the watch section consists of junior Aerographer's Mates who plot the weather data on our maps and later trace our analyses on the maps prior to transmission. We, Ableman and I, analyze the surface and several upper-air levels on acetates with grease pencils. After I approve of his analyses we work together to produce the required weather and ocean forecasts. We are working against transmission deadlines. We work: 12 hours on, 12 off, 12 on, 48 off, 12 on, 12 off, 12 on, 72 off.

Those of us who are also pilots have to accomplish our monthly and annual flight requirements (day, night, instrument and check flights) during our "off time."

One morning while I am giving the FWC command's morning weather brief, a few months after my arrival, either the Operations Officer or the Executive Officer is disagreeing with my local forecast, and in general giving me a hard time. Finally having had enough, after finding a piece of chalk, I write equations that cover our present situation on the board and begin explaining which of the terms should dominate, and why. After finishing, no one questions me. Upon glancing at Captain William (Bill) Kotch, the Commanding Officer (CO) of FWC/JTWC, he is smiling at me. Captain Kotch always briefs the Admiral, Commander, Naval Forces, Marianas Islands (COMNAVMAR), and his staff after our morning brief.

In December Captain Kotch selects me to attend the special Advanced Tropical Meteorology course at the University of Hawaii starting in January. This course is for people who already have a degree in Meteorology. Students take this one class, lecture and lab, all day for about two months. Most of the students are in the military. Ron Hughes attends from FWC Pearl, he was a classmate of mine at NPS, but he was in the Bachelors program. The Meteorology Department Head gives most of the lectures. My work earns an "A."

Upon my return, the Operations Officer, Commander Selfridge, reassigns me to my watch section to continue as FDO. My section is very glad to have me back! Even my communications personnel welcome me back – not sure what happened during my absence.

In late April we have a typhoon in the Western Pacific that was not anticipated. Most of the Typhoon Duty Officers (TDOs) are on leave in Japan. Being the only FDO who volunteers to assist, I am standing both FDO and TDO watches and working 16 hours on and 8 off. The Director of the Joint Typhoon Warning Center (JTWC), Air Force Lieutenant Colonel Hutchinson, is working around the clock, with a few hours off now and then. Both Captain Kotch and Lt. Col. Hutchinson are very grateful for my help! JTWC is a joint command (Air Force and Navy), with Captain Kotch commanding both FWC & JTWC.

Captain Kotch has been strongly urging me to request a change in primary designator from Operational Aviator (1310) to Meteorologist (1530 at this time, later 1810). Considering: 1) I have now missed two tours in an operational aviator billet, 2) during the deployment of the USS *Hornet* I missed getting the normal amount of carrier landings for a cruise because our VAW-11 Detachment KILO (among others) had been off-loaded in Japan and 3) the comments of the three meteorological officers designated 1310 at NAS North Island - I reluctantly apply for change in designator.

Later, after requesting to become a TDO for JTWC, Captain Kotch confers with Lt. Col. Hutchinson and then allows me to switch. During my stay on Guam, no other forecaster worked for both units. The amount of work may be slightly less with JTWC, but the stress is a lot higher! One has a lot of calls from operational units who do not like your forecast and are trying to get you to modify it.

Our forth child is born in the Agana Naval Hospital during mid August 1962.

I am standing TDO watches while Typhoon Karen is approaching Guam. We, the other TDOs, the Director and our CO, have agreed it will cross directly over Guam. I take a call from the General's Office of the Strategic Air Command (SAC) in the states. He tells me that I am making a mistake; no one away from Guam thinks it will cross Guam, and wants to know the basis of my forecast! He is referring to the reported tendency of forecasters to forecast strong storms to pass over their locations – do not know if this is true or just what others like to believe. I have been called because SAC does not want to fly their bombers away from their base on the northern end of Guam. The forecast is not changed! The Air Force finally flies their bombers from Guam, just barely in time to prevent damage!

I am off duty when Typhoon Karen (Category 5) crosses over Guam and we spend the evening and night in our officer's quarters on Nimitz Hill. Fortunately we had been told our quarters were designed to withstand the forces associated with this storm – and we believed them – this turned out not to be true. The noise of the storm sounds like a continuous freight train rumbling through our drive. Our wooden "typhoon-proof" shutters are closed, but the rain comes right through in horizontal streams. Jackie and I keep mopping up the floors with towels and wringing them out into buckets, which we then empty into the toilet. We hear gravel from adjacent roofs hitting our car in the car port, which is just outside our bedrooms.

When the eye of Karen finally passes directly over Nimitz Hill we enjoy a brief respite. Against Jackie's wishes I step outside to have a quick look around – our son's fort is gone. The whole experience has been horrific – and it is only half over!

The terrific noise and battering of the storm resumes as the wall-cloud of the eye passes over us again... Jackie and I are soon exhausted, and can't keep up with the water coming into our quarters. Our hands ache from the wringing of the towels. We finally all huddle together in our smallest bedroom. With our new baby in her basket, our 2 year old in her crib, Jackie with our 4 and 6 year old children on the single bed and me in an easy chair from the living room ride out the remainder of the typhoon. Believe it or not – we fall asleep in spite of all that is going on around us!

During the night of the devastation the watch section of FWC was huddled in a secure vault in the basement of the main building. JTWC personnel were not present; typhoon forecasting responsibilities had been passed to our backup Navy and Air Force units in Japan.

In the morning all is quiet and when we look out, the trees are bare - the leaves are gone! The jungle behind our house is now brown stocks – which allow us to see a great distance! All the quarters on Nimitz Hill are without electricity and water for over two weeks. Only one of the quarters on this hill is damaged – part of the roof is missing. We all pitch in to help one another. FWC/JTWC has an emergency generator and limited freezer space is available for members of this command. Also, we obtain emergency bottled water at FWC/JTWC. Groups of families share the outside cooking duties at various quarters on Nimitz Hill.

Lt. Col. Hutchinson, CAPT Kotch and visitor. (By author)

Our secondary building, mainly used by JTWC, which is shown behind the officers, is written off after the devastation. However, our main "Quonset Hut" (shown at right) is made operational again. Our sign is found after the storm a good distance away, and is temporarily reinstalled. The main building is replaced in 1965 by a typhoon-resistant building.

Typical Guamanian house, business and Coast Guard hangar at Agana airfield (By Author)

Since FWC/JTWC is not operational, one of my duties is to canvas the island and record damage and heights of the ocean surge generated by the typhoon. During this investigation, I learn the navy officer's quarters on Nimitz Hill were designed for 150 mph sustained winds, not the 150 kts of Karen!

Fortunately, they were over designed, over built, or both!

The barometric trace of Typhoon Karen recorded at FWC/JTWC indicates the lowest pressure is about 935 mb. 1013.25 mb is standard sea level pressure. Karen had sustained winds of 150 kts (173 mph) and gusts to about 175 kts (over 200 mph). (Today this would be classed as a "Category 5" Typhoon. Note: mb is millibars of pressure and kts is nautical miles per hour.)

Unfortunately, an Air Force Lockheed Super Constellation, C-121G, loaded with emergency supplies for Guam, crashes on 4 December 1962 near the top of Nimitz Hill during a night approach to NAS Agana. All crewmembers are killed; the plane and most of the supplies are a total loss.

One night while working on another typhoon forecast, I have almost decided this typhoon should just miss going over Okinawa as it passes. A military person on Okinawa calls and reports he has the typhoon on radar and begins giving me positions of the eye. The plotting of these positions clearly indicates that the typhoon is going to cross Okinawa. Fortunately, Captain Roper, an Air Force Captain who is in his second year at JTWC, has given me a lot of good tips during my training. One of them is that one can *never* take radar reports from a non-meteorologist at face value. I use my best judgment, assume his reports are in error, and keep the forecast typhoon just off the coast as it passes by. Later I am praised for not believing these reports! During the subsequent verification, which is completed for all forecast cyclones, it is confirmed that the typhoon had not crossed over Okinawa.

Around the first of the year Captain Kotch informs me that my request for change in designator has been approved. He is very happy! I am still a naval aviator, but will never again serve in an operational flying billet.

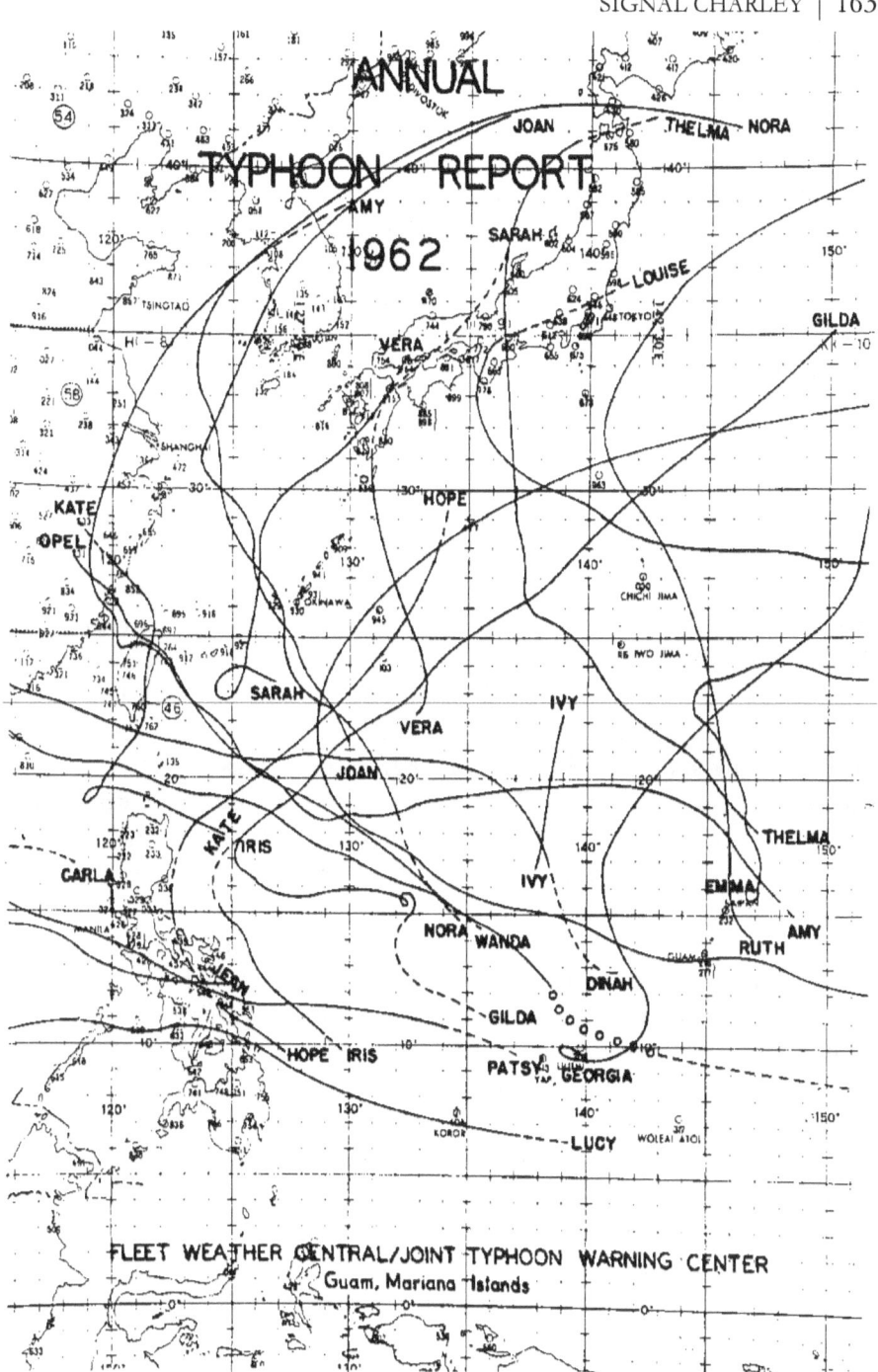

ANNUAL
TYPHOON REPORT
1962

FLEET WEATHER CENTRAL/JOINT TYPHOON WARNING CENTER
Guam, Mariana Islands

During the 1962 season we have 24 named tropical cyclones in the western Pacific, one of the most prolific seasons. (As shown on the cover of the Annual Typhoon Report 1962 for FWC/JTWC, on previous page).

Around the first of May, when it is time to assemble those TDOs who will be there during the new tropical-forecast year, I return to being a FDO for FWC to complete my tour in July. However, transportation back to the states is not available till August. We return on a Navy R6D-1 (DC-6) four-engine transport to Travis Air Force Base, California via Oahu, Hawaii where we have a brief rest stop.

Flying

While stationed on Guam the only plane available for me to fly (as first pilot) is again the SNB-5. On a morning flight, shortly after takeoff I am visually inspecting the plane, which is standard procedure. As I look over the port engine I see oil streaming from it. I immediately declare an emergency and land. I give the Maintenance Chief a hard time after I find out he has jury rigged a repair of the oil line – and had *hoped* it would hold. But his biggest error is that this maintenance is not reported on the yellow sheet. ALL repairs that have been performed on a plane should be listed on the yellow sheets for that aircraft, with proper signatures, for the pilot to see before flight. Also, certain maintenance requires a test flight rather than a regular flight. This type of repair probably should have had a test flight.

After a normal flight in an SNB I make a good landing and during the rollout, while the tail is still in the air, suddenly the whole instrument panel (containing both pilot and copilot instruments) falls off! My copilot grabs it the best he can, to keep it off the yoke. After I park the plane I am told that maintenance people had been working on the panel and had not buttoned it up properly. Again there is no indication of this work on the yellow sheets! Why the panel had stayed up while taxing to the duty runway, during takeoff, throughout the flight and even on initial touch down is a real mystery.

After I'd been flying about 3 hours or so, on another flight, I return to Guam's Navy/civilian field to land. As usual, the Navy tower reports the direction and speed of the wind, and then clears me to land. There is a strong left crosswind and I am unable to get the plane set-up properly for landing, so I tell the tower I am taking it around for another try.

"Roger," reports the tower. Again when I call in before commencing the approach they give me the direction and force of the wind. I am doing much better this time, but as I look towards the Navy tower I see a lot of people standing on the elevated decks of the main building and tower watching me land. I think it is strange, but I have my hands full with this strong left crosswind. To line-up on the runway with a crosswind, you have your windward wing down and opposite rudder applied – in this case my left wing is down and my right rudder is pushed in to keep the nose heading down the runway. Just before the left wing tip hits the ground I rapidly level my wings and feed in left rudder. Even on the run-out the plane has to be carefully handled to keep it on the runway. After I have parked the plane back at the line, I am informed that everyone was watching because I was attempting to land at the maximum allowed crosswind conditions for an SNB – 25 kts at 90 degrees to the runway!

While in Hawaii, I fly the SNB-5 with Ron Hughes (a classmate at NPS who is now stationed at FWC Pearl and is also attending this special course). He ensures I get a good look at Oahu and the islands to the east-southeast, but we never fly to the Big Island (Hawaii). We get a good look at Haleakala National Park on Maui. The volcano is very impressive, glad the weather is good on this flight. We also fly the same path that the Japanese planes flew over Scholfield Barracks on their way to attack ships in Pearl Harbor.

The most interesting flight I take on Guam is when I fly a Navy Captain, who sits in the copilot's seat, to the Mariana Islands Tinian and Saipan. I have forgotten if he had been a pilot or not, he is commanding officer of a ship that has come to Guam. But he had been to these islands during WW II; and wants to see them again.

It is my first and last time to land on Tinian. There is no control tower; the runway is left over from WW II. Tinian is where the two B29s that carried the atomic bombs to Japan were based. There are two plaques marking the parking places of the two planes.

We fly on to Saipan, which I have landed on before, though not often. As we walk off into the jungle, he keeps pointing out where guns had been and where Japanese women had committed suicide, some while holding their babies in their arms, as they jumped from cliffs, etc. When we come upon a large and deep cement containment basin, he explains that the Japanese had developed a new type of cement that could hold water even when it is "green" (not properly cured). They built the basin to provide water for the forthcoming large build-up of Japanese soldiers. The Japanese had kept the basin camouflaged. He thought that it was very likely that Amelia Earhart had been killed on Saipan after she had seen the large cement water basin – to keep the basin a secret.

We are deep into another part of the jungle, when he says, "No; we need to go further to our left." All of a sudden we are standing in front of a camouflaged entrance to an underground hospital. We carefully walk past a swarm of mosquitoes, and go down the stairs to the underground facility – which is spotless! How, he found this entrance is beyond me. He is very appreciative of my taking him to the two islands, and I enjoy hearing all his stories and getting to see the B29 plaques.

Just a little east of due north of Guam is the small island named Rota. It is the southernmost island of the Mariana Islands. As I understand it, Rota was never invaded by the US during WW II. However, there was a Japanese gun emplacement that had to be avoided (until they ran out of ammunition) when our planes were flying from Guam during WW II. The Japanese also had some fighter planes there – but they were all destroyed, either in the air or on the ground. Supposedly, the Japanese surrendered to the natives after they ran out of ammunition and could no longer fight. The natives flew a large white flag when this happened.

While stationed on Guam I fly to Rota infrequently to either take or bring back some official. There is no aircraft control tower on Rota. The first time I land on Rota I am a copilot for a pilot who has been here before – as were all other new pilots who would later fly this mission.

I remember that the jungle has so overgrown the ends of this Japanese fighter strip that before you get back into the plane, you walk from the Rota passenger terminal (a wooden shed) to the takeoff end of the strip, while inspecting the condition of the composition for support of the plane and tramping down plants growing through the surface to give yourself another 10 to 20 feet of usable surface. Therefore, upon positioning the plane for takeoff, you all but put the tail of the SNB into the jungle to maximize your takeoff run in order to improve your chances of clearing the tall trees at the other end of the runway.

It is very intimidating looking from the cockpit of the SNB at those palm trees at the not so far end of the old fighter strip– the peacetime Navy doesn't usually fly from jungle strips, especially in low powered aircraft. We advance the throttles as far as we dare, holding the plane with its brakes, with full up elevator to keep the tail wheel on the ground (your nose off the ground!!!). Upon release of the brakes we go to full throttle on both engines – simultaneously, of course. This is a maximum performance takeoff for a taildragger. You retract the main gear immediately upon liftoff. We never seem to clear the trees with much room to spare.

On one scheduled flight to Rota, I am told that I will be flying by myself (only pilot – which is very unusual) and will have a passenger that needs to do some work over Rota. I think they mean *on* Rota. While I am inspecting the plane I see that the door of the plane has been removed. The Chief tells me not to worry about it, because it has to be off for my passenger to do his job. I notice a lot of "boxes" in the plane, where the rear seats belong, but the Chief tells me that the cardboard boxes are very lightweight and the weight-and-balance of the plane is still within limits.

SNB/TC45 (National Museum of Naval Aviation)
Note door just aft of third passenger side window.

My passenger sits in the first passenger seat behind the copilot's seat for takeoff. As we approach Rota we see that the central part of the island, including the ridge, is obscured in clouds. My passenger *then* informs me that he has to drop sterilized mosquitoes while *flying* over Rota.

He tells me that he is conducting an experiment to reduce the population of mosquitoes on Rota. And that we have to be very low and slow for him to safely drop his mosquitoes. That he has to cover most of the island with his "fixed" mosquitoes. I point to the clouds – yes, according to him, we have to fly into them! I try to explain to him that it would be better to do this on a clear day. He says he has been waiting a long time, and he is out of time. It has to be done today!

Fortunately I know the topography of Rota fairly well. So I make passes from the clear areas towards the high ridge, once in the clouds I time the flight before I do a maximum climbing turn to avoid flying into the ground. We make run after run!

I finally look back to see how he is doing and if most of the boxes are gone – it really scares me!

He is standing in the open doorway with some sort of strap attached to him as he flings the contents of the boxes into the slipstream. He really has a *lot* of nerve!

As I understand it, several years later this same experiment is conducted on one of the islands of Hawaii. I never do learn for which US agency my passenger is working.

What is the most fun I had in an airplane after leaving squadron VAW-11? Answer: A Navy Douglas JD-1 Invader from a utility squadron based at NAS Atsugi, Japan arrived on Guam to provide target towing for gunnery practice by a destroyer squadron. I notice on the air station bulletin board that the JD-1 has an empty copilot seat. I sign up for it and later get to fly in it as copilot. At this time I did not know only the pilot has flight controls – yoke and rudder pedals.

I think it is very interesting the way the tow procedures are handled. No, it does not bother me having ships firing at the tow – as a Midshipman I had seen the other end of this gun firing practice from a destroyer and a battleship.

The fun is the extracurricular flying he does before we land. I am really impressed with the performance of this plane!

Note: During WW II the Army Air Corps designation of this plane was A-26 Invader, during the Korean War and Vietnam its Air Force designation was B-26 Invader – both the Martin B-26 Marauder and the classification of "Attack" were stricken by the Air Force after WW II. This plane has a wing span of 70 feet, 20 more than a Skyraider, and length of 51.25 feet, about 12 feet longer than a Skyraider. The A-26 was powered by two Pratt and Whitney Double Wasp engines, rated at 2,000 hp each, by the end of WW II.

For fun, one day I catch a ride in an Albatross (a twin-engine amphibian), as two navy pilots practice "splash-and-goes" in Apra Harbor. We takeoff and land at NAS Agana.

On 5 August 1962 I get an acquaintance of mine, LCDR Kidd, who is a plane commander in squadron WV2 to take me on a Typhoon penetration mission in support of JTWC. I am really looking forward to seeing a typhoon from the inside. However, we lose an engine (we have three left) on the way out, and have to turn around and return to Guam

5-hours later. This plane is the Navy version of the Super Connie with radar domes on its back and belly.

I am not in the cockpit when we lose the engine and did not know we had a problem, until I ask why we have turned around. It is odd looking out and seeing the propeller stationary – this is my first and, fortunately, last experience to see this.

Another time when our (Kidd and mine) schedules allow me to fly is on 18 March 1963. This is a recon flight for FWC/JTWC – which I have arranged. We fly south, looking for the ITCZ (Intertropical Convergence Zone). We cross into the Southern Hemisphere on this flight (this is my first and last time to be in the Southern Hemisphere), but never find the ITCZ. This is a 5.6-hour flight.

13

FWC SUITLAND – PROJECT FAMOS

My next orders are to report to Fleet Weather Central (FWC) Suitland, in Maryland, for duties in Project FAMOS. I do not know what FAMOS stands for, nor do I think my orders are unusual. However, when reporting for duty in late August 1963 to the Commanding Officer (CO) of FWC Suitland, Captain James West, I can tell something in his greeting is not right.

I then report to Lieutenant Commander (LCDR) Harvey Jenkins, Officer-in-Charge (O-in-C) of Project FAMOS (Fleet Applications Meteorological Observations Satellites). Both FWC and FAMOS are in adjacent spaces in Federal Office Building (FOB) #4 in Suitland. Most of the occupants in this building work for the Weather Bureau.

The world's first polar-orbiting weather satellite, the Television Infrared Observation Satellite (TIROS), was launched on 1 April 1960. Subsequent TIROS's (II through X) are launched through 2 July 1965.

Project FAMOS was just formed to determine the operational usefulness of the TIROS data. Harvey and I are the only people working in Project FAMOS. Harvey informs me he was passed over for promotion to Commander last year and if he does not make Commander this time he is going to resign. He also confirms that the CO of FWC Suitland had assigned him to this position. I am not impressed with him! He is not selected for promotion and he does resign. This leaves me, as a senior Lieutenant, filling a Commander's billet.

By this time, I have heard that another Lieutenant within FWC had been previously selected, but not assigned, by the CO to my position in FAMOS. Under normal conditions, COs fill all the positions of their commands, except the Executive Officer (XO). Therefore, Officer's orders rarely specify a specific billet within a command, below the level of XO. Thus, my orders to Project FAMOS are *unusual*; but then Project

FAMOS is very unusual! Almost no COs have a special project, with separate funding, attached to their commands.

After meeting a lot of meteorologists in the Washington D.C. area, including Vince Oliver – the leading satellite meteorologist of the Weather Bureau – I set out to perform my job as best as I can.

Soon a senior civilian working for the Headquarters of the Naval Weather Service informs me the funding for Project FAMOS has been approved and transferred for *my* use. Now, knowing the Headquarters is paying for Project FAMOS, I assume that Captain Kotch selected me to come to Project FAMOS. He had left Guam about two or three months before we did to become the new Commander of the Naval Weather Service.

Upon discovering that FAMOS should have an Aerographer's Mate First Class (AG1) assigned, I have a talk with the CO, and shortly Fred Volpe, AG1, is assigned. After learning how to hire civil service workers, I hire a secretary, Jackie Welsh, and a Meteorologist, Fred Bittner, from the Weather Bureau.

This is truly cutting edge work. Now, it is hard to realize that this is the first time the navy has really tried to match synoptic scale cloud cover with analyzed weather maps. The observed cloud cover certainly does not match many of our preconceived cloud models, which had been developed over the years based upon numerous land reports.

The first TIROS had been launched while I was attending the Naval Postgraduate School. None of our professors, especially the Department Head, Professor Duthie, were impressed and nothing useful was anticipated from seeing the tops of clouds. On Guam when the TIROS images were available, which was seldom and not in a timely manner, only JTWC would try to use them to post locate tropical cyclones and estimate their prior maturity.

We work with the leading meteorologists of the Weather Bureau and NASA who are trying to do the same work: to ascertain the meaning of what we are seeing – types of clouds, implied temperature differences, wind shear, and etc. We are trying to determine what type of structure is being represented and what changes to this structure can be inferred for the near future. Also, how is dynamics, such as positive vorticity advection, being represented in the cloud patterns?

We finish the first paper within the Navy on how to interpret and use this data operationally. Not having the authority to directly transmit this paper to the FWCs and other interested commands, I write a draft cover letter for our CO. Assuming he will change this draft letter as he sees fit prior to sending out FAMOS's first report to the fleet.

When called to his office I cannot believe his comments and position! He says he cannot send it out because he cannot tell other commanding officers what to do! I then request he send it to the Commander of the Naval Weather Service, for transmittal. He says he cannot do this because he has not been tasked to prepare this report. Finally, he agrees to send out our first report if we change *all* our definitive findings on how to operationally use this data to suggested uses!

Later, upon receiving a call from the Headquarters of the Naval Weather Service on why this report is not more declarative on the use of this data, I explain what has happened with my CO. The response is for me to write exactly how this data should be used operationally. I never have any more problems in getting the operational guidance reports sent by the CO. Do not know who talked with him, but it sure worked. Of course I'm sure this did not make him happy. He never is friendly towards me.

We need some research accomplished that we cannot do, so I learn how to write a Request for Proposal (RFP), get it issued, properly evaluate the responses to the RFP and get a contract issued to the company that wrote the wining proposal.

In preparation for the launch of the TIROS VIII satellite, which will carry the Automatic Picture Transmission (APT) camera system for the first time, in addition to the standard wide angel lens used on prior satellites. APT allows any ground station with an APT antenna to receive in real time three satellite-based pictures per orbit. Each picture covers 800 miles on a side.

I report to the CO of USS *Saratoga* (CVA-60) in Mayport, Florida for temporary additional FAMOS duties in October 1963 because she will be the first carrier to have an APT antenna installed. I am here to assist in the planning for the installation of this antenna and associated equipment.

I return to *Saratoga* on 17 December to be aboard during the first test receipt and the evaluation of this antenna. TIROS VIII launches on 23 December. We prove the concept, there are no electromagnetic interference of the signal or with other equipment; but the antenna, which must track the satellite, is too large for operational use on a carrier, and is removed shortly afterwards at the behest of the ship's CO.

During this second visit I happen to run into Gordian Stevens, a fellow AD-1 student pilot in our flight section at Cabinass Field (third from the right in our flight section photo). He is no longer flying, his back was broken during a "cold" cat shot and he could not pass the flight physical after this accident. I have forgotten which job he had aboard ship; but it had nothing to do with the APT system.

The success of the APT system makes the work of Project FAMOS even more relevant and necessary for our weather centrals and facilities around the globe to improve their forecasting capabilities.

I finally find another meteorologist, Ralf Nielsen, who I think is qualified to help us. He is a civil servant working as a meteorologist for the Air Force.

Fortunately, LCDR Bill Arnold is ordered in as the O-in-C of Project FAMOS about 10 months after my arrival. I have been Acting O-in-C because an officer who is two ranks junior to the *billet rank* can't be CO or O-in-C.

Within a few months Bill is promoted to Commander (CDR) and I am promoted to LCDR. By the way, I knew Bill because he had been the Aerology Officer, as a Lieutenant, aboard USS *Hornet* (CVA-12) when I was flying from her in VAW-11 Detachment KILO. Aerology is the former name for meteorology, and the navy retained this name for many years. In fact, rated enlisted meteorology personnel are still Aerographer's Mates.

One of my accomplishments is to convince Captain Paul Wolf, the CO of Fleet Numerical Weather Central (FNWC), that we (FAMOS personnel) can provide derived data from our satellite observations on a routine operational basis that the FNWC numerical analysis program can use in sparse conventional data regions to improve its analyses. I had been told over and over that Wolf would never allow satellite-derived data into his analysis programs. After flying to Monterey I am really grilled by him in his office, but I answer all his questions and explain how we will derive this data, etc. Just as I think he will say, "No" and send me from his office, he says he will allow a trial run.

Fred Bittner and I then take turns generating the promised satellite-derived data and transmit it from FWC Suitland to FNWC for their operational use. FAMOS is still providing this data when I depart in June of 1966.

At the beginning of my last year at FWC Suitland, I request my next billet to be a meteorologist aboard an aircraft carrier – this is a billet that is strongly desired (even required?) for promotion to Captain. However, when notified that I have been selected to fill the Stockholm, Sweden billet; I cannot say no! Again, I think Admiral Kotch, as Commander of the Naval Weather Service, is influencing my orders. (*Historical Note:* Admiral William Kotch is the first active duty meteorologist to be promoted to Admiral.)

Flying

One time I am flying some passengers in the UC45J (designation for SNB-5 – thanks to Secretary of Defense Robert McNamara) from Andrews AFB outside Washington D.C. to a small civilian field in New Jersey near New York City. I am used to flying left hand approaches at fields I know, but this New Jersey field is new to me and it has a right hand approach for the duty runway, plus a crosswind. On final I never quite get the plane set right for landing, so I call the tower and tell them I am taking it around. On the second approach I have it set-up right and make a good landing. So I am very miffed when one of the passengers (a navy civilian) expresses how unhappy he is that I had taken it around – apparently it had scared him.

Hey, if you pay nothing and arrive in one piece, what more do you want?

On another flight, I am flying copilot with Bill Arnold as first pilot in a UC45J heading from Andrews AFB to a field in Maine to accumulate our required (monthly and annual) flight time, and to buy some lobster. We are about 1/3 of the way, Bill is flying on instruments, and I am keeping an eye on the plane. I notice ice starting to build on the leading edge of our wings. Icing conditions have not been forecast for our flight, so I am surprised to see the ice. This is the first flight that I have experienced icing.

The UC45J has deicing equipment on the plane - for the propellers, leading edges of the wings, and leading edges of the stabilizer and fins. The UC45J has two rudders. The leading edge of each wing has an inflatable boot that will break the ice loose. But one has to wait until there is enough ice on the wings so it can be broken off.

I know that Bill has come from the Fleet Weather Facility Kodiak (Alaska); so I assume he has often flown in icing conditions. I keep my cool and watch the ice build up on the wings. Finally, when I think we have enough accumulation to take action, I mention the ice on the wings to Bill.

To my big surprise he says, "What do you recommend?" I find out later that Bill also had never flown in icing conditions before!

I work the wing boots, and also turn on the propeller deicing – an alcohol mixture that coats the blades/ice. We can hear the ice hitting our plane as it is slung from the propellers. It is very interesting. It does not last very long, we run through about three or four cycles of buildups and clearings before we are clear of icing. This is also my last encounter with icing on a plane in flight.

Bill and I routinely swap places, first pilot and copilot, on our flights. On the cross countries that involve landing and parking, we swap places before returning. Bill is a rank senior to me, I make Lieutenant Commander the same year he makes Commander – we have our wetting down party together.

On another flight to Maine for some lobster, I am flying as first pilot and he is the copilot. The first pilot normally takes off and lands the plane. When we get to the field in Maine, I land the plane – several times as it turns out. I do not have the right attitude when I touch down, so we bounce and I have to add power and re-land, etc. I've forgotten how many times, I think it is three "landings" before I get it to stay down.

I have managed to not get into "porpoising" – which progressively gets worse and often leads to an accident. However, it is the worst landing I had ever made. Fortunately, Bill is calm and lets me handle it! If two pilots try to land a plane at the same time it usually quickly becomes a much bigger problem.

I taxi in and park the plane. Bill seems to be in no hurry to get out of his seat, so I get up and head to the tail, where the exit is. At the door I wait to let him go before me, since he is senior. He motions for me to go ahead, and I again try to defer to him. He says, "Go ahead." So I do

Later I am talking with him about the landing, etc. and I ask him why he had basically insisted that I step out of the plane first. He says the custom (I assume this is from true multiengine type planes – patrol, etc.) is for the pilot who lands the plane to exit first. He does not want others at the field to assume he has made this terrible landing!

I start having some minor problems landing the UC45J, so I decide flying (landing) a tricycle landing gear plane would be easier. I start transitioning to these types of planes, in addition to flying the UC45J, as early as November 1963. I fly as copilot of the U11A. This is a smaller, more modern utility aircraft than the UC45J; it only carries two pilots and two passengers. When we have a passenger, it is always an Admiral. We take him to his destination and return him, that day or another. The person I fly with works for this Admiral in the pentagon

In January 1966 I get more serious, and complete two familiarization (FAM) hops in the US-2B. This is the utility version of the S-2F – a twin-engine carrier ASW plane. I make 14 landings on these two hops, with different instructor pilots. The second flight is also a check ride, which I pass. Now I can be scheduled to fly in this plane, which I usually do with Walt Mitchell, another meteorologist at FWC Suitland and who had flown the S-2F operationally.

The bad part of flying various versions of the S-2 is that after the plane entered service the Navy discovered that if an engine is lost on takeoff or landing the pilots often cannot put enough force on the rudder peddles to control the aircraft in this emergency situation.

Therefore, the plane is equipped with a power boost on the rudder, only to be used during takeoffs and landings.

The problem then becomes that occasionally this power boost system will activate itself, giving full right or left rudder during takeoff and landing – which means the plane *always* crashes!

So, at the beginning of each flight the pilot and copilot decide which malfunction they want to chance – engine failure or power boost failure! We always choose engine failure, and turn off the power-boost system.

In March I know I am being ordered to Stockholm, Sweden for my next assignment. So I inquire which type airplane is being flown by flight personnel attached to the embassy. This embassy is an Air Force embassy – meaning the highest-ranking officer is from the Air Force. It is the T-29 – a version of a twin-engine civilian passenger plane built by Convair. In the Air Force it is mainly used to teach air navigation to student navigators.

The navy in the Washington D. C. area flies from the Andrews Air Force Base. So I go to the Air Force side of the field and request flight training in their T-29B. There is not enough time and room for me to complete the full pilot training, but I talk them into qualifying me as copilot, during the first part of April. So, after a 1.2-hour daylight flight and a 4.0-hour day into night flight with me making 4 landings, I have my copilot check ride. I fly in the pilot's seat for 3.5 hours. This flight includes simulated instrument time with 2 instrument approaches and 5 landings. I am now copilot qualified, by the Air Force.

The rest of my flying, four flights, is in the US-2B at Andrews for April and May 1966.

14

SWEDEN: PROJECTS OFFICER, UPPER ATMOSPHERE RESEARCH PROGRAM

University of Stockholm

We travel from New York City to Southampton, England aboard the SS *United States*.

Upon arrival on Tuesday, 7 June 1966, I promptly report to the Office of Naval Research (ONR), Branch Office, London as the new Projects Officer, Upper Atmosphere Research Program, Stockholm, Sweden. The Commanding Officer (CO) of this ONR Branch Office is my CO while I am on this tour of duty, plus I will be attached to the Embassy in Stockholm.

While I am being briefed on my duties and responsibilities, Jackie and our four children, shown above, see some of the sites of London.

On Saturday, 11 June, we fly to Stockholm, Sweden and at the airport we are met by Commander Glenn Hamilton, the person I am relieving, and some embassy personnel. At the American Embassy on Monday I report aboard to the Naval Attaché. This is an "Air Force" Embassy, meaning the Air Force Attaché is the senior military person assigned to this embassy. Therefore, the embassy aircraft is an Air Force plane rather than a Navy plane.

Tuesday, Glenn escorts me while I check-in with Professor Bert Bolin, Head of the Institute of Meteorology of the University of Stockholm and formally register as a student of the University at their administrative offices, several blocks away.

At the institute I share an office with a Swedish student. The work is hard, and is performed one major subject area at a time. The six major areas are: mathematical methods in meteorology, thermodynamics and cloud physics, atmospheric chemistry, atmospheric radiation, dynamical meteorology and the general circulation of the atmosphere. After reading the required books and published papers for a given course, you ask the Professor in charge of this area to schedule your exam. The exam is an all day, 7 to 10 hours, affair – the first part is taking a written test, and the second part is oral, which may be practical and/or hypothetical questions asked by the professor. I always pass each exam, but some of my fellow students, including my roommate, do not. He always says, "I will hold my thumb for you!" Which is the Swedish way of wishing a fellow good luck - and it always works for me!

In addition to studying we attend special lectures by the professors and visiting scientists.

While working on my degree at the University of Stockholm, I continue to spread the word on the use of satellite data. I brief Swedish Air Force meteorologists on the interpretation and operational use of this data.

After the World Meteorological Organization (WMO) announces a conference on "Applications of Satellite Data" to be held in Moscow, Russia, I am strongly encouraged to attend this meeting to make further contacts with foreign meteorologists and evaluate the extent to which the Russians are using meteorological satellite data operationally. In particular, I am to determine if the Russians are obtaining our APT satellite data with their antennas and if so, are they using this data operationally.

Professor Bolin is pleased to have me attend the conference as a student of the Institute. Therefore, I will not be attending as a navy meteorologist. At this time it is unusual for US military personnel, who are not assigned to the Embassy in Moscow, to travel to Russia. The Naval Attaché of the American Embassy in Moscow is notified by message from our embassy of the dates I will be in Moscow, which makes me less apprehensive about the trip.

Since I must obtain a visa from the Russian Embassy in Stockholm, and they are knowledgeable about everyone attached to the American Embassy, they know they are allowing a naval officer to travel to Moscow. Thus, this is *not* a clandestine visit.

As I board the Aeroflot Airliner in Stockholm, I can tell by the nose of the aircraft that this plane has been converted from a bomber to an airliner; and may be just as easily converted back to being a bomber.

I have no problems passing through Immigration at the Moscow airport. At the hotel I am informed that I will have to share a room with other WMO attendees. I am placed in a room where a Scotsman and an Englishman are already sharing. I am given a cot to sleep on in a sitting room separate from the bedroom, which has two single beds. So this is sort of a small suite. Neither of the two gentlemen knew each other, and both were surprised that they would be sharing a room. Fortunately, they do not mind my sharing their spaces with them.

When I board the special bus for the WMO conference, which is being held in the Russian Weather Bureau's operational office building in Moscow; I am surprised to see Vince Oliver. He is more surprised to see me; we had worked together in Suitland, Maryland. I had helped him teach several satellite interpretation courses to groups of international meteorologists – from Iceland and western European countries. I learn he is a guest speaker at this conference.

During a break in the presentations, which are for the most part being spoken in the native language of the participants with simultaneous translations being provided via headphones, we tour the complex. I am surprised to see large, outdated computers, running with vacuum tubes. The building seems run down. Most people, including the Russians, will not ride the elevator up to the auditorium. We hear that the elevator often works improperly or not at all. And that this is typical of elevators in buildings not normally visited by tourists.

The information being provided by the Russian meteorologists is theoretical rather than practical; for the most part the other presentations are useful for operational meteorologists, with enough theory to provide the background for why the particular application is valid.

One afternoon, when operational applications are being discussed in several smaller rooms at once, the one I have chosen to attend does not have a presenter. The person in charge asks several times if there is anyone present who knows enough about this type of satellite interpretation to give an off the cuff lecture. Since no one steps forward,

I finally volunteer. After I finish, the person in charge thanks me for helping out and I get a round of applause from the participants.

Each night a group of English-speaking members go out together to be entertained. We go to the Moscow Circus, the Swan Lake Ballet and an opera (sorry, I don't remember the name). However, during intermission at the opera I notice the line for ice cream, seemingly popular refreshment, is moving very slowly. Upon further inspection I see the reason, each cone is being weighed and then reweighed after the contents have been changed (increased or decreased) to ensure every person is receiving *exactly* the same amount of ice cream! Is this Communism run amuck?

One noon, rather than returning to the hotel for lunch, four of us decide we will eat at a Russian cafeteria near the office building to mix with typical Russian people. To our surprise, one stands in line to pay for the food before you get it. We join the end of the line and hope for the best. When we arrive at the clerk's stand we see there are no pictures to which we can point. Fortunately, a gentleman (KBG?) behind us speaks enough English to help us out. The clerk hands each of us a slip of paper that lists what we have selected and paid for. At each position along the display of food the server takes our slip and loads our plate with what we have bought. The only silverware available for use to withdraw from the bins is a fork and a spoon.

A little after we have seated ourselves at an empty table and started eating, a person comes from the back with knives for each of us to use. He is speaking Russian, so we are not sure what he is saying, but he appears to be treating us as honored guests. The smiling people seem to be very glad that we have chosen to eat with them!

During the last break in the conference, a person whom I had recognized the first day as being a student in one of the satellite interpretation classes I had helped teach in Suitland, very nervously approaches me and asks in a low voice, "Are you the naval officer who taught me in Suitland?" When I respond with yes, his eyes get big and he quickly leaves me – with what he thinks is a big story to tell when he returns to Iceland.

The conference ends at noon; but our hosts have arranged a trip to the local space museum, for those who are interested. Of course I go, and it is impressive!

Some surprises: When checking into the hotel one must surrender one's passport - seeing tanks moving on the streets of Moscow – Russians must have a government permit for travel before boarding a train, plane, etc. - men lining up outside after work to drink beer from a state run sidewalk corner distributor; they have several mugs, but they are not washed between drinkers - getting to go into the Kremlin museum - seeing Lenin's Tomb, you see him encased in glass lying dressed on a slab of marble (the guards insist we cut in line near the head of a very long line of Russians) - how smooth, quiet and timely the subway system is. As a Midshipman, the Paris subway had impressed me, but this Russian system seems better, and each under ground station is beautifully decorated, no two stations are the same!

During my last morning in Moscow, I bundle-up all the material that I have gathered and walk over to the American Embassy. I am surprised to see an armed Russian guard at the gate in the fence around the Embassy! I assume the best approach is to ignore him, which I do and have no problems. In the embassy I show the Marine at the desk my ID card and request to see the Naval Attaché. The Marine is surprised when he sees my ID card. Soon a person comes out front and escorts me into the Attaché's Office. He is upset that a naval officer is in Russia without his knowledge. I am surprised that he had not seen the message sent to him about my trip. He gladly takes my material to be sent to me at the Embassy in Stockholm via diplomatic pouch.

Both the Scotsman and the Englishman have checked out while I have been gone. While I am waiting in the room until my departure time, a hotel employee comes in and looks around and then claims to see damage on the small table between the single beds. He informs me I will have to pay for it! I inform him that I slept on the cot in the other room and deserve a discount for not having a regular bed to sleep on!

While I am checking out at the desk in the lobby, they tell me I must pay for the damage in the room. I keep telling them that I did not do the damage, that I was sleeping in another room on a cot and deserve a discount! When they retrieve my passport; they stop asking for extra payment. I am traveling with an Official Passport – red colored – that states I am "Abroad on an official assignment for the United States government." There is no indication that I am in the military.

At the airport, I see several Swedish attendees who will be on the same SAS flight as mine to Stockholm. One of them I recognize as a Swedish military meteorologist. Again having no problems going through Immigration Control, I am relieved! However, it is short lived; for when the passengers leave the building and walk out to board the plane, they must walk between two rows of armed troops eyeing each person, in an intimidating way, as they walk to the steps leading up to the plane.

I relax once we are airborne heading for Finland.

After we land in Helsinki, those going on are detained aboard the aircraft until those getting off have been processed by Immigration. Then everyone aboard is ushered off the plane in order to pass through Immigration before getting right back onto the plane. I never had to do this at any other airport, so I do not know whether this is normal for this airport or if something special is occurring at this time.

When I get off the plane in Stockholm, I am glad to be home!

After I am home a few days, Jackie tells me that I am acting very depressed. I tell her that I had never thought there would be a war between Russia and the US because the Russians had too much to lose – but after visiting the country, I didn't see that the Russians have that much to lose!

Later, the West German weather service attendee contacts me to present a briefing on the utilization of satellite data for the West German weather service. I comply with this request.

Through the Embassy the US Air Force requests that I provide a briefing at a base within West Germany for their meteorologists. So I make another trip to Germany to comply.

About a month before what is now called the "Six Day War," I am requested by the Israeli attendee at the WMO conference to present a one-day applications course for their operational meteorologists in Israel. I accept, and the tentative date would have placed me in Israel just before the start of this war. I have forgotten why the date is changed, but I arrive just after the war has been concluded. I am met at the Tel Aviv airport by my contact and he escorts me to the head of the line for the Immigration Office. I no more than hold up my Israeli Visa and my passport than I am waved through. I never do learn exactly where my contact works and whether he is civilian or military.

He drives me to my hotel and stays with me while I check-in and walks me up to my room to make sure everything is all right. He will pick me up at 0900 and take me to the headquarters building for my presentation.

At 0900 sharp he picks me up and drives me to our destination. I am well received, and then my talk begins with my slide show. Then it gets down to very detailed questions related to the local area. I tell them what I know, and they share with me their experiences and shower me with many questions. Some of their questions I cannot answer because of the regional conditions; but I do tell them how I would approach their problems. It is late in the afternoon before the discussions are over. Oh, yes, we did break for a quick lunch. They thank me for coming and say they have learned how to make better interpretations and understand the associated dynamical considerations. I have also learned a lot about their local problems.

Next morning he picks me up to go site seeing. We drive as far north as Haifa, where our Embassy is located, and then southeast to Jerusalem, where we have lunch. It is very moving for me to be in the towns of the Bible, and see the mixture of various religions in a small place. The old town of Jerusalem appears to be as it must have been centuries ago!

We drive east to the Dead Sea, where I go swimming.

In the evening I am invited to have the evening meal with his family, which I really enjoy because it is a typical meal. I hope I didn't ask too many questions; they are very gracious and also ask me questions about the States.

The next morning he picks me up and takes me to the airport, and whisks me through Immigration again. The officials seem to know him.

I hope I have helped the Israeli meteorologists and in exchange I have had a once in a lifetime experience.

The second summer a few of us students fly up to Kiruna in Lapland, above the Arctic Circle, to participate in the launch of a meteorological research rocket at the Swedish rocket facilities. It is very interesting, and even exciting, being in the blockhouse looking through small, reinforced portholes during the count down and launch of this rocket!

The next morning, rather than driving my rental car, I ride with an American couple to the facility to see and listen to the evaluation of the raw research data from the rocket. This couple, whose husband has a PhD in meteorology, has been visiting the Meteorological Institute and has been invited to attend this launch.

He is driving a little fast, but not excessively so, but he still manages to miss a left-hand curve on the access road and hits a tree, just off the pavement, head-on. Just before the accident, I had nearly grabbed the wheel to go straight off into the big yard of this farmhouse – sure wish I had!

After the accident, I see the driver, hunched over the wheel, not moving and his wife in the back seat looks dead. So I climb out the now windowless area of the right side door and crawl a few feet away from the car before passing out.

I hear voices and see ambulance attendants. I hear that they think I'm in the worse shape and should be taken to the hospital first. While they are preparing to put me on a stretcher, I see this shoe close to my face – and because of some odd compulsion I reach out my hand and pull on the end of one the laces, which untie his shoe!

In the ambulance they keep talking to me, trying to keep me conscious I think!

In the hospital, a doctor sews up my face below my left eye; where I had taken out the rear view mirror. They place me on a side table in the x-ray room and leave me alone for a while. Then the whole room starts whirling around, and I hang on to the edges of the table for dear life. I close my eyes so I can't see the room whirling, but I still feel very dizzy!

Their x-rays indicate that three "wing-bones" of my lower spine, two on one side and one on the other, are fractured. (Later it is theorized that the wife's knees have broken the bones in my back as she is thrown forward during the accident. The rear of my seat is lifted as it is torn from its railings, and when it comes down to really cuts up her left foot. She is admitted to this hospital, but her husband does not require hospitalization.)

Late in the afternoon or early evening Jackie receives a call from Assistant Naval Attaché Commander Bobby Inman that I am in the hospital for injuries received in an automobile accident. (Yes, this is the same Bobby Inman, who became an Admiral and then Deputy Director of the CIA.)

The next morning Jackie calls me to see how I am doing and asks if I would like for her to fly up and help me. I respond, no, that it is more important for her to take care of our children; I am getting better and should be home in about a week.

In a few days she receives my luggage containing my bloody jacket and other clothes from the accident in addition to all but one set of clothes I'd need to return home. Needless to say this is a big shock to her to see all my bloody clothes!

On the way to the airport in Kiruna I ask the cab driver if he knows where a car that was recently wrecked might be. He has read of the car accident in the newspaper and takes me by the wrecking yard where the car is. After I see the car next to the fence, I am glad to be alive! The roof of the car had buckled just above where my head would have been in a normal sitting position! In fact I would not have been surprised if I had been told all three occupants of the car had been killed.

The cab driver tells me that when there is an accident on that road it is always with that tree. Officials have tried numerous times to get the owner to let them cut it down; but he always refuses.

When I arrive at the Stockholm airport, I am the last person off the plane. I have stiffened up while sitting during the trip and I cannot walk as well as I had in the hospital. I can barely put one foot in front of the other! After what seems like an hour I finally approach the terminal. Looking down from a balcony is Jackie and our four children, with tears in my eyes I smile up at them!

Much later, in the Air Force Hospital in Wiesbaden, Germany, their x-rays indicate that five of these little "wing-bones" of my spine have been broken, three on one side and two on the other. Between two of the breaks on one side, the calcium has grown and fused together – forming a "double backbone."

My required research is esoteric. Professor Bert Bolin, Head of the Institute of Meteorology of the University, requests that I determine the significance of atmospheric radiation in the Omega Equation. At this time, the Omega Equation is very important in numerical forecasting models.

The work for a Filosofie Licentiate, the degree I earn, is harder and covers the whole scope of meteorology rather than a major area as is done in the States for a Doctors of Science degree in Meteorology. As you know, in the States, a DS degree in Meteorology, or whatever discipline, is basically a PhD in that discipline *without* the research. However, my research is much less than would be required for a PhD in Meteorology. At this time in Sweden, one has to establish oneself within the field of meteorology with numerous publications over many years and submit a compilation of your work to be reviewed to obtain a Swedish Doctors degree in Meteorology, as a few other countries are also still doing.

Upon leaving Sweden I am too senior to be a meteorologist aboard a carrier. My orders are to report aboard the Naval Postgraduate School in Monterey, California to teach meteorology.

We fly from Stockholm, Sweden to Oslo, Norway and then to New York City. The last leg is aboard a PAN AM jet aircraft. This is first time for Jackie and children to fly aboard a jet aircraft. Not sure if the plane is a Boeing 707 or a Douglas DC-8.

Flying

The embassy plane is a VT-29B – which means the front of the plane has seats for passengers and the back is bare for carrying cargo. In addition to the self-contained folding stairs for boarding on the right side of the plane (just behind the cockpit), there is a cargo door near the rear on the left side (just aft of the last window).

Air Force T29 modified (VT29B) for embassy service. (US Air Force)

After I report in to the Naval Attaché, I check in with the Air Force Attaché. They are glad to have another pilot. On my two flights in June 1966, I fly local flights with Lt. Col. Stewart, for 7 hours. I make 3 landings each flight and 2 instrument approaches on the first flight and 16 on my second flight. They want me to qualify as Command or First Pilot, and I say no thank you! I do not have the time to learn all the systems and procedures of the T-29 to qualify as First Pilot! And I do not want to be a pilot in command without this knowledge. This does not make them happy!

In July we fly over to an airfield in Finland to pick-up children of families working for our embassy in Finland for a visit in Sweden. Our return clearance is to climb on our heading for Stockholm to an altitude of 5,000 feet. Lt. Col. Stewart is doing the flying. When he starts climbing right through 5,000 feet, I grab my yoke and shove it forward to level us off at 5,000 feet. As I do, I remind him we have only been cleared to climb to 5,000 feet. I am not impressed with his concentration skills! He is not impressed with my copilot diplomacy. Not having gone to copilot charm school – I do not know exactly what I am supposed to have done! He gets mad, but never says what I should have done. I think he is embarrassed.

On 14 August I am flying with Col. Foote, the Air Force Attaché, (who arrived early in July), on a trip to Wiesbaden, Germany. During

WW II he flew twin-engine bombers and remembers attacking Wiesbaden.

Among the passengers are Jackie and our four children, Stephen, Christy, Laurie and Nancy. During the takeoff run we hit a bird with our left propeller or engine. We have to decide what to do – at this time; I wished my family were not aboard! In the past, I never minded making decisions for myself and/or my crew; but I did not like this situation!

Everything seems normal, so we continue the flight to the airfield just outside Wiesbaden, with no problems.

Jackie and the children got to fly on the embassy plane because they have medical appointments at the Wiesbaden Air Force Hospital. I also have my annual flight physical. It is a different eye test than the Navy gives; and for this reason or others I fail the eye exam! However, on my records they mark that I have passed. This obviously disturbs me, in several ways.

A number of years later, I learn that before my examination a senior naval aviator from Rota, Spain was grounded after his examination at this hospital. The Navy and the Air Force got into a big hassle. The end result is that the Air Force at this hospital will not ground any navy flier – regardless!

The highly questionable "missions" I flew as copilot of the embassy plane I have chosen not to comment upon. So this is the end of this flying section.

In the first part of January 1967, for various reasons, I submit a letter to be removed from flight status. My request is granted; so I have made my last flight as a naval aviator on 12 December 1966, with about 1,760 flight hours and 70 carrier landings on six carriers. All but one of the carrier landings were arrested landings – I had one bolter, which is a carrier landing that did not catch an arresting wire.

15
NAVAL POSTGRADUATE SCHOOL

Assistant Professor of Meteorology

In July of 1969, while reporting back aboard the Naval Postgraduate School (NPS), my promotion to Commander is confirmed. Professor George Haltiner, Head of the Meteorology Department, selected me from the officers offered him by the detailer. He had been my thesis advisor when I was a MS student. After the University of Stockholm confirms my graduate work, the Bureau of Naval Personnel evaluates my academic work and Swedish degree and officially credits me with having earned a Doctors Degree. Following this, thanks to the efforts of Professor Haltiner, I am made an Assistant Professor of Meteorology. At this time no other military person at NPS has academic rank.

While in Project FAMOS I had flown several times to Monterey to present briefs at NPS on the operational use of satellite data. I am now very pleased to learn the Meteorology Department has incorporated the use of satellite data within the curriculum.

I am assigned to teach introductory courses of meteorology, which normally would have been taught by Professor Haltiner, or grudgingly by other professors – who naturally prefer to teach within their specialty. I am also assigned to teach some of the lab courses, which my fellow military Instructors are teaching.

Since I had helped teach satellite interpretation to classes in Suitland, Maryland and then to various interested groups while stationed in Sweden, I feel I can handle teaching these courses. But I am nervous my first time. However, I soon get into the swing of things. No one in the department ever monitors any of my teaching. I am not an easy teacher; that is I do not teach just the highlights, even if these are introductory courses. I really want them to understand the underlying principles – without this I feel they will have trouble later on.

At the end of each term students anonymously fill out evaluation sheets. In general, in each course my evaluations fall into two approximately equal groups, those that think I am excellent and those that think I am poor – I receive few marks in the middle of the comment choices. I never understand this.

However, I am always pleased when my old students come by to thank me for giving them a strong foundation. Several of these students tell me they answered questions on tests in later courses by what I had taught them, rather than what had been presented in their respective classes!

On the multiple-choice questions of my tests that require calculations, all my answers are possible if one makes typical math errors. (Divides rather than multiplies, etc.) Thus, one cannot work until he/she finds a matching answer because all the other answers are not reasonably possible to obtain.

Also I would sometimes have "None of the above, because_____." And yes, one time, I give a student credit for marking this because of the way he had interpreted my question. So I do try to be fair about it!

Having access to the school's computers, I teach myself to write software programs in FORTRAN. When Professor Haltiner learns of this, he assigns me to teach the meteorology course where the use of the computer is introduced to the students. This is an interesting course to teach. I am not teaching them FORTRAN, but I am teaching them how one can use a computer to solve meteorological problems.

After convincing Professor Haltiner to let me contact other Curricular Officers about having their students take a general course in meteorology, I find one who is willing to let me try. I develop a special course for this purpose. The majority of my students find this new knowledge useful, but unfortunately their section leader (a Lieutenant Commander, who was a poor student) did not like the course. Not surprisingly, the Curricular Officer does not pursue it. Well, I tried!

During March 1972 I volunteer to take the place of an oceanographic student who cannot make the scheduled 15-day trip to the Navy Arctic Research Laboratory (NARL), Barrow, Alaska for hands-on fieldwork. Since most of this work is accomplished outside, where one is not allowed to work alone, there must be an even number of participants; in this case, 8. I am senior to the Oceanographic Instructor, CDR Karl

Schriner; but we both agree that he is in charge of this expedition. After other professors and instructors agree to cover for my absence, I am permitted to assist the oceanographers.

On Saturday, 4 March, we depart from Monterey for Point Barrow, Alaska via NAS Whidbey Island and NAS Kodiak in a four-engine Navy R5D (DC4). When we land at Whidbey to spend the night, Karl discovers there has been a mix-up at the BOQ for our stay, and we can't spend the night here. To my surprise, he turns to me to get it resolved. I have the Chief, who is handling the BOQ berthing; put the air station Duty Officer on the phone. This officer tries to shrug it off, as mix-ups happen; but when I threaten to have the Admiral of NPS call his Commanding Officer, who is a Captain; everything is suddenly straightened out! At this time I also check on our return accommodations.

After we land at the NARL field, on 6 March, the plane taxies inside their hangar before we depart our craft. This is the first time I had ever seen or heard of a plane taxiing into a hangar. As each of us climbs down from the plane, we are issued cold weather gear to wear over our uniforms – boots, pants, jacket and a fur cap. Which we immediately put on!

During our stay, we fly the NARL twin-engine Navy R4D (DC3) out to where the Project AJAX scientists are living on an ice flow in the Arctic Ocean. We land and after off-loading a few treats, we split-up and talk with the scientists in their work tents to learn about the research each group is performing. While doing this we keep an ear cocked for the sound of our aircraft engines. It is so cold that the engines must be kept running to keep warm. If one quits we are stuck on this floating block of ice! Of course they tell us this after we have gotten off the plane.

When it was time for us to leave, we use Jet Assisted Takeoff (JATO) bottles to depart the ice floe, as shown on the previous page. But our main skies were attached to our main wheels; rather than replacing them, as in the above picture. (US Navy Official)

The whole trip is very interesting and informative, and I am glad that I have been allowed to be an oceanography "student."

I am an advisor for a few MS theses. After responding to a call for scientific papers by the American Meteorology Society, I am invited to present the results of one of these theses at the Fourth National Conference on Weather Analysis and Forecasting held in May 1972 at Portland, Oregon.

I strongly try to get orders for a career-enhancing billet following my NPS tour – as the Seventh Fleet Meteorologist - but the detailer does not agree with my choice. So I select to go to Fleet Numerical Weather Central (largely located on the NPS campus, at this time) since it means we do not need to move and this tour should be good for my career, but not as good as a Fleet Meteorologist – which is on an admiral's staff.

16

FLEET NUMERICAL WEATHER CENTRAL

Computer Systems Department Head

In June of 1972 I report aboard Fleet Numerical Weather Central and am assigned Head of the Computer Systems Department by the Commanding Officer, Captain Sam Houston. My people supervise and operate all the hardware within the computer center (military personnel) and maintain the operating systems on the various computers (civil servants).

We had met Captain Houston in the Washington, DC area while he was serving in the Headquarters of the Naval Weather Service, under Admiral/Captain Kotch. He had the ONR Sweden billet before Glenn Hamilton (the person I relieved) and had encourage me to take the Sweden billet. I had given a briefing for him to senior Air Force officers at the Pentagon.

Once I get my feet on the ground, I inform Captain Houston that I *have* to secure the computer center. As it is, anyone attached to the command can walk into the building where our two main and communications computers are running. Not only is this a security risk, some of our programs and their outputs are classified, but it is disrupting for my people. He says he will go along with my request, if I can get the Head of the Development Department to agree.

This means I have to get Leo Clarke to go along with my idea. (Yes, this is the same Clarke who interviewed me when I came to NPS as a student.) So I have a meeting with him and he is very reluctant to agree. Finally I suggest having a meeting with his personnel, where I will explain how my people will be able to satisfy their requirements.

I convince most of his Developmental personnel, who write and maintain most of the operational software running on the mainframes, that their requirements of fast turnaround of the output from the programs under development and the diagnostics from the operational programs will be met.

From that time forward the computer center at Fleet Numerical has been operating in a secure environment.

The next item on my agenda is stopping the development of new programs being written in assembly language – about one step up from machine language. Technically this is none of my business, but I cannot tolerate inefficiency and unnecessary job security for civil servants!

The first CO of Fleet Numerical, Captain Wolf, insisted *all* software must be written in assembly language – that is until he had a deadline to meet on some ASW (Antisubmarine Warfare) support software. An officer informed him he could meet the deadline, *if* he could program in FORTRAN. Captain Wolf called his bluff and the officer made good on his boast! From then on, FORTRAN is allowed, but not required. Early FORTRAN compilers could not write as tight a code as an experience assembly language programmer. This is the excuse used for not writing new programs in FORTRAN, by certain individuals.

But the capabilities of the FORTRAN compliers kept improving and the size of computer memory kept increasing with time. Ted Hess is a senior civilian who screens all submissions to either change or install new operational programs. However, he refuses to keep up with the times and always programs in assembly language.

One time when Ted Hess is on leave, I explain to Captain Houston all the advantages of programming in FORTRAN and all the disadvantages of programming in assembly language. He agrees with my facts and logic, and declares *all* new programs must be written in FORTRAN.

When Ted returns from leave and discovers that new software must be in FORTRAN, he comes storming into my office – he knew I was the one behind the change – to tell me what he thought of me going behind his back! This is not the first or last time I had Ted mad at me.

I have no direct meteorology duty, so, in my spare time, I write a program to bogus a 12-hour forecast field - which is the "first guess" field for the next analysis. If the first guess field for a numerical analysis is flawed the resulting analysis will also be flawed. Ted Hess, who is responsible for the analysis software, has been trying for days to remove an erroneous low on the surface near the South Pole. Learning of this upon returning from leave, I laugh and show him what my software will do. He is both appalled by my approach and glad for the results! My program becomes operational, but it is to be used only in emergencies by the Duty Officer.

Our two operational Control Data Corporation (CDC) 6500 computers, named Bonnie and Clyde, are the first operational computers in the US, if not the world, who can "talk" with each other! This is accomplished by having a large bay of Extended Core Storage (ECS) to provide the same meteorological data to both computers simultaneously, provide extra work space, and a communications area. During the running of the global forecast model, each computer is working in a different latitude band. But they must sync-up at the end of each band. Therefore, when a computer finishes with its band, it checks to see if a message has been sent by the other computer, if not; it says, "I'm waiting for you." Likewise, when the other computer finishes, it checks for the "waiting" message; if found, it sends, "Ready for new data." When both computers are "Ready" to start the next band, the new meteorological data is loaded into ECS for both computers, and they are informed to commence the next bands. This continues until the whole globe has been forecast for 12 hours; then the whole thing starts over for the next 12-hour forecast of the globe. Once the forecast limit is reached, 72 hours at this time, the computers are ready to work independently of each other on the next programs in the operational run.

My biggest challenge and accomplishment is to keep the operational computational run on schedule while we move all the computers from our computer center on the NPS grounds to our new building on the grounds of the old Monterey Naval Air Station. For years other departments of Fleet Numerical have been working in converted wooden barracks across the street from where our new building is being built.

We complete this move by obtaining another CDC 6500 to install in the new building before we commence the move. This computer is named Samson. We use a pick-up truck to transport magnetic tapes between the buildings – very high data rate transfer at that! We cut a few corners, but never deviate from the operational output schedule at any time – we get a big "Well done" from the Commander, Naval Weather Service!

When we briefly (12 or 24 hours) had only one mainframe running in each building we resort to our emergency one computer global model – this model is not as accurate as our two-computer model; but we have no other choices.

After the move, we have another CDC 6500 computer for development of new software; which was sorely needed before the move, and as a back up to Bonnie and Clyde.

During my last year at Fleet Numerical, the new Commander of the Naval Weather Service, Captain Sam Houston, offers me the Executive Officer (XO) position at the Naval Environment Prediction Research Facility (NEPRF). This seems good to me, I am moving up the responsibility ladder from Department Head to XO and not having to move from Monterey - NEPRF is just across the street from Fleet Numerical.

17
NAVAL ENVIRONMENTAL PREDICTION RESEARCH FACILITY

Executive Officer

In June of 1975 I am ordered to the Naval Environmental Prediction Research Facility (NEPRF) in Monterey as the Executive Officer (XO). Again this starts out being mainly management and administrative rather than meteorology. The XO of any command does what the Commanding Officer (CO) wants him to do – this means the XO does what the CO does not want to do himself!

The senior civilian position, Director of Research, is filled by an oceanographer who is also the Department Head of Oceanography. Unfortunately, NEPRF receives an order from Headquarters that as of 1 October 1975 the oceanography support being provided by NEPRF will no longer be funded. Captain Cody Sherar, our CO, gives me the task to determine how this will be implemented.

Some of our oceanographers are willing to perform meteorology duties and be absorbed into other departments, where funding is available. However, when our Director of Research discovers that he will not be permitted to do any oceanographic work, he decides to leave. So NEPRF advertises the position of Director of Research as a meteorologist position.

When the Director leaves, I am made Acting Director of Research. In this position, I become involved in the writing and/or evaluating Requests for Proposals (RFPs) for the research work we cannot accomplish in house. I then learn the details of getting these RFPs issued by the Supply Department on the naval base in Alameda, the responses to the RFPs properly evaluated in accordance with government guidelines and obtaining signed contracts. I also learn the procedures for monitoring the progress of the contractor and the evaluation techniques of final reports and deliverables.

During this reorganization, I am requested by the CO to do the required work to have a civil servant, who is classified as a meteorologist, removed for poor performance. This requires over six months, going through all the required paper work and civil service hearings, to be fired! Both my legal training in San Diego and the knowledge I have gained by programming scientific programs on the Fleet Numerical computers, which NEPRF scientists' use, comes in handy. He has a PhD in astronomy, and after being fired he becomes a local taxi driver while looking for other work in the Monterey area.

The CO appoints me Project Manager of Satellite-data-processing-system Software Development Office (SSDO). Satellite data, when correctly processed, is becoming an important data source for numerical models for various environmental applications. Again, this is cutting edge research and development.

Captain Sherar is a former naval aviator and is a member of the Navy Flying Club of Monterey. When we travel together in California on official business he always rents a plane from the Navy Flying Club so we can fly to our destination – such as Point Mugo and Palm Springs. The small plane is always a high-winged, single-engine with dual controls and side-by-side seating. The plane also has a back seat, which at times has a passenger. He shares the flying with me, once we are at altitude and on course for the next reporting point. It sure has a very light feel on the controls!

On one trip to Palm Springs we, the CO and I, meet with Dr. Krick to discuss some possible research his company might be able to perform for us. He is a meteorologist well known for his statistical approach to forecasting. He is also the only meteorologist on General Eisenhower's staff who told the General that it would be safe for D-day to be on 6 June because the number of soldiers killed due to less than ideal weather conditions will be less than those killed when the Germans are expecting an invasion.

The CO and I review all the applicants that have applied to become the new Director. We select the top five candidates, and I am sent to go interview each of them. When I return, I give a full report of the strengths and weaknesses of the candidates.

The CO decides the new incoming CO, Captain Pete Pettit, should make the final choice of Director.

After Captain Pettit becomes the new CO, about the end of July 1976, my number one choice for Director is invited to come to Monterey for an interview. The CO selects this candidate, Alan Weinstein, to become our new Director of Research, and he accepts.

Captain Pettit later decides to fire a civil servant, who has a PhD in meteorology, during his probationary period for poor performance. After I complete the paper work for the probationary person, and he is fired; he sues me in Federal Court for $50,000. I go to the Federal Court in San Francisco and talk with a lawyer in the Justice Department about how to handle this. The court, without a formal hearing, dismisses the suit because of a technicality. It has something to do with the fact that I am a legal resident of Texas not California.

In my year group there are two of us with a doctors degree; but neither of us are selected for promotion to Captain – also neither of us have been a meteorologist aboard a carrier. So I put out a few feelers for employment and my best offer is to remain in Monterey.

I select Sunday, 31 July 1977, for retirement from the Navy.

Author's retirement picture

18
CIVILIAN MANAGER/
SENIOR SCIENTIFIC ANALYST

Program Manager

The Vice President of Systems and Applied Sciences Corporation (SASC) of Maryland, Sharad Tak, is the gentleman who recruits me to work for SASC. He occasionally visits Fleet Numerical Weather Central (FNWC) and Naval Environmental Prediction Research Facility (NEPRF) to maintain contact and learn of possible contractual work for his company. The CO of NEPRF always has me deal with him. On one of these visits he approaches me as I am running up the steps of FNWC, and asks if it is true that I am planning to retire. I briefly respond yes, and then hurry on to take care of whatever I am going to do at FNWC. Later that day or the next, he again spots me outside and asks me if I would be interested in opening an office for SASC, on the Monterey peninsula. I tell him yes, and he requests I send him my resume. It would have been inappropriate for him to talk to me, or me to talk to him, about possible employment while in my office at NEPRF.

After reviewing my resume he sends a formal offer for employment, dated 8 June 1977. After considering all my offers, I respond with an acceptance dated 21 June. My acceptance contains a few conditions and understandings, to which Sharad Tak agrees.

As the Program Manager of the Space Applications Division of SASC for the west coast, I locate office space in Monterey and commence business on Monday, 1 August 1977. I have one person, Keith Nuttall, working for me. We have two contracts with FNWC obtained by Shard Tak to begin our work.

My previous work dealing with the issuance of Request for Proposals (RFPs), evaluating them and issuing contracts prepared me to respond in a winning manner to the RFPs issued by FNWC, NEPRF and other organizations outside the Monterey area. We respond well and win most of the contracts we bid on. I locate a good headhunter and we are soon employing 10 people. We buy a computer to be used by our secretary, who at one time had been a computer operator, and buy numerous terminals for my people to use in our office rather than having to go to the bull pen at Fleet Numerical to develop software.

Our most important contract, as measured in terms of savings for the Navy, is the Optimum Path Aircraft Routing System (OPARS). This is a set of programs that determines the route (if not specified), altitudes (allowed, depending upon direction of flight) and speed to be flown for the minimum consumption of fuel between departure and destination (which may be the same point if the route is specified). All the planes in the database will be jet or turbo-prop powered aircraft.

Another company had previously won the contract for developing the algorithm – we had somehow not seen the RFP for this. But then this company does not win any of the four or five follow-on contracts to develop the total operational system. We win three of them.

According to our evaluation, their design, as provided by the Government (FNWC) cannot be implemented in the real world. Unfortunately, the originating company of the design will not assist us, even though they have been given a special contract by FNWC to provide assistance. This makes the Fleet Numerical Contract Monitor upset, but for some reason the contract is not written in such a way as to force them to reveal "company secrets," which they are now claiming is their right.

So Bud Hinson and I have a long talk on how to develop a workable design for the minimum fuel consumption routing algorithm. He was a submariner rather than an aviator, so I explain the details of flight planning. Bud goes to the NPS library and returns with several ideas to consider. We talk them through and come up with a couple of approaches to try. He quickly implements a design that appears to work for our test cases.

Another big problem I have (I'm doing the programming) is getting our meteorology databases small enough to be practical on the computers of Fleet Numerical. We need to have special regional meteorology databases for our programs to use in different parts of the world to minimize both the computer memory space and the execution time of the programs. The problem is I must fit *four* parameters into a 60-bit word while retaining 20 bits of accuracy for each parameter. Our parameters are: pressure, temperature and two wind components (u – east west and v – north south).

One evening at home while mulling over this problem, the light dawns! I remember from my aviator days that planes flying above 18,000 feet fly flight levels – which are based upon the standard atmosphere; that is, with a fixed sea level pressure of 29.92 inches of mercury. Therefore, flight pressure may be determined from the flight level and temperature. All the planes in the aircraft database (expressed in terms of aircraft performance data - which we are also programming) will be flying at flight levels. Therefore, the initial design provided by the Government requiring all these parameters is *wrong*!

That is, wrong for most of the flight. The only time the actual pressure might be needed is for flight at and below 18,000 feet: takeoffs and departures, and approaches and landings. But the difference between the calculated fuel used for these two parts of the flight are *insignificant* between doing it correctly, with the actual pressures; and approximately, by using the pressures of the standard atmosphere.

Fortunately, I am able to convince the Contract Monitor, John Garthner (a former naval aviator), of this insignificance! Thus, our meteorology database, can be composed of temperature and winds (each with 20-bit accuracy) packed into a 60-bit word.

We do such a great job on developing and implementing OPARS that we also win *all* the yearly software maintenance and update contracts (increasing the variety of aircraft in our data base and adding additional operational flight options) until years later when Fleet Numerical personnel finally take over these functions themselves. In addition, a steadily increasing number of Air Force pilots and navigators begin using OPARS rather than the Air Force's flight planning program.

The yearly savings in cost of aircraft fuel for the Navy is significant; in fact it is *greater* than the total annual cost of operating Fleet Numerical! Thus, my OPARS work is the most significant work I did for the Navy, when measured in monetary terms. Though, I think the breakthrough work I did with the satellite data interpretations should be in first place – it laid a strong foundation to be built upon over the years.

Another initial contract that pays-off well for us is the work I do to develop a new contouring program for Fleet Numerical.

One day I began wondering how the various contoured products are produced at Fleet Numerical. I certainly understood how one accomplished this manually – we learned this at NPS and I practiced it enough at FWC/JTWC Guam!

So the next time I am at Fleet Numerical I stop by Jim Long's office to have a chat about this subject with him. He is responsible for the maintenance of the operational program that produces all the contouring by the computers at Fleet Numerical. Initially another software company in Monterey developed this program, but Jim has made a lot of enhancements to the basic program. Jim has been a friend of mine since I was the Computer Systems Department Head of Fleet Numerical.

He gives me a copy of an Air Force publication containing several software applications. One of them is on a new way of improving contouring. I take it with me and in the evenings I begin studying the improved concept. As is usual with this type of article there are some gaps, at least for me, on just how certain aspects can be accomplished. However, I finally figure it out enough, with some suggestions from Jim, to start writing a contouring program on my home computer.

I finally develop a way of using cubic splines on a basic four-by-four sub-grid (of the original grid-point data) to produce the desired contours. The speed and accuracy can be changed based upon the amount of interpolated points desired from the splines.

Soon a RFP arrives in the mail at work requesting the development of a faster and more accurate contouring program for Fleet Numerical. My accidental timing and work could not have been better!

We win the RFP, and develop the initial program that meets all the program size, speed and accuracy requirements specified in the RFP. Over the years we receive numerous other contouring contracts to add various modifications to the basic software.

In addition, the use and application of cubic splines pays dividends on other contracts.

The RFPs from NEPRF are research orientated, while FNWC's are operational. We receive a NEPRF RFP which is requesting research, but it also specifies the approach to be followed. I want to do the work, that is personally to do the work, but I strongly disagree with the specified approach. So I submit a response that fully explains how I will follow this specified approach; but I also include why I think this approach is not the best approach and explain how I think the research should be accomplished.

Following company policy I submit a copy of this proposal to our home office at the same time as I submit the proposal to the Navy Procurement Office. In a few days my boss in Maryland called to inform me, in a nice way, that I should *never* tell the customer that he is wrong. He is convinced that we will not win, and thus I have wasted time and money by responding in this manner!

When we receive notification that we have won the bid, I promptly go to NEPRF to speak with the person for whom the work is to be performed – the contract monitor. I am completely surprised, when he tells me we won because he thinks we are the best qualified to do the work, *but* he still wants the approach to be done his way. I had thought that if we had won it would be done my way.

Over the years we win contracts to perform work for other facilities and companies. Among these are: the Air Force Geophysics Laboratory near Boston; the Air Force Global Weather Center near Omaha; the Naval Oceanographic Research Center in Mississippi; the Naval Pacific Missile Test Center at Point Mugo; the Naval Weapons Center, China Lake, California; the NOAA Marine Operations Center, Pacific in Seattle in conjunction with the Scripps Institute of Oceanography at La Jolla; the Ocean Routes Company in San Jose, the NASA Ames Research Center at Moffett Field (where I have employees working on site) and the Jet Propulsion Laboratory (JPL) in Pasadena.

JPL later sends me a parchment to acknowledge my work, which reads:

THE PROOF-OF-CONCEPT VOYAGE OF THE FIRST OCEANOGRAPHIC SATELLITE *SEA SAT* GRATEFULLY ACKNOWLEDGES THE DEDICATED SUPPORT OF HARRY D. HAMILTON IN THE DEMONSTRATION OF THE CAPABILITIES OF MICROWAVE REMOTE SENSING OF THE WORLDS OCEANS

Applied Numerical Solutions

A few years after I have opened the SASC Monterey Office, the company splits in two. Vice President Sharad Tak takes approximately one-half of the company and names it SASC Technologies, Inc., with himself as President. The Monterey Office is included in this new company. Later this company is renamed STX. The ST is for Sharad Tak and the "X" because it is an "In" letter these days.

During this reorganization I felt a backup plan for income would be prudent, so I form my own company, Applied Numerical Solutions – with the catchy phrase, ANS is the Answer. So for a couple of years or so I often work 7 days a week, and long hours each day. I always do all my ANS work on my own time. However, after awhile I become confident that I can go back to running my own company if required; so I stop doing this extra work.

Senior Scientific Analyst

Both Fleet Numerical and NEPRF, who are our major sources of work for our Monterey people, stop issuing RFPs for software support/research work, and start using onsite software specialists through contracts administered by the Government Services Agency (GSA). Martin Marietta (MM) has the first GSA contract for the Western Region of the US. I contact the local MM manager in Pacific Grove to see if he has any openings for NEPRF or Fleet Numerical. He has an opening working for Leo Clark, Head of the Development Department – both Meteorology and Oceanography – of Fleet Numerical Meteorology and Oceanography Center (FNMOC). (Yes, this is the same Leo Clark who had interviewed me as a prospective MS student and had been the civilian head of this department when I was the Head of the Computer Systems Department. The name of Fleet Numerical changes every now and then; but it is the same organization – that is why I usually write Fleet Numerical.)

After Leo informs the local MM manager that he would be pleased to have me as a GSA specialist, I submit my application to MM and my two weeks notice of departure to STX. In reality, I work four weeks or so for STX in order to have a smooth transition in June of 1989 from STX to onsite GSA contractor. (The problem is I am personally doing some software work for Fleet Numerical at STX that Leo wants me to finish before I come to work on site; and STX is in no hurry for me to leave.)

I really enjoy these onsite years. Not having to submit responses to RFPs and being concerned with getting and keeping enough work for my employees means I can now devote all my attention to the application of numerical meteorology and oceanography.

In particular, I am directly supporting the work of Charlie Mauck, Head of the Tropical Division. My main work is implementing new and maintaining old tropical cyclone forecasting computer programs. On my own, but with Charlie's support, I take over *all* the software dealing with tropical support of the Joint Typhoon Warning Center (JTWC). (Yes, it is a small world). Charlie is very surprised that I am able to get the cooperation required to accomplish this. However, Fleet Numerical people are normally glad to give up responsibility for programs that are not within their main interest. In this way I become responsible for all of the software that processes the incoming tropical cyclone forecast requests and the outgoing forecasts. I write interactive software for the Operations Dept. personnel to process incoming requests. In addition I rewrite most of the existing tropical cyclone forecast programs and implement new forecast programs to support JTWC.

All the tropical models, except one, are based upon statistical correlations. The exception is the One-way influence Tropical Cyclone Model (OTCM), a truly primitive equation regional model. Note: in meteorology it took a long time before computers were large enough and fast enough to run global primitive equation models. Primitive means the whole equations without approximations – the true basic equations. Thus, in numerical meteorology primitive is good, not bad! OTCM became operational about 1977 for the western North Pacific Area; so it is showing its age. However, it is the best forecasting model we have for JTWC, based upon the yearly average errors. But, for any given cyclone forecast, the Typhoon Duty Officer of JTWC does not know which model will be the most accurate. Stated another way, he must still be a meteorologist!

I am responsible for the software that provides the Navy Operational Global Atmospheric Prediction System (NOGAPS) with bogus tropical cyclone data at the locations and with the wind speeds determined by the tropical forecast centers, JTWC and the National Hurricane Centers in Florida and Hawaii. These synthetic (theoretical) observations represent the circulation of each tropical cyclone (in a warning status) in terms of conventional data and are added to the synoptic data stream; but are flagged as synthetic for the model assimilation software.

In about 1996 Fleet Numerical receives the GFDL mesoscale regional model from the developers at NOAA's Geophysical Fluid Dynamics Laboratory (GFDL) on the campus of Princeton University. This model is being run by the National Weather Service to support the National Hurricane Forecast Centers in Florida and Hawaii. Mary Alice Rennick, who has gone from a fellow contractor (working across the hall from me when I first became an onsite contractor in Building 4) to a civil servant, had made arrangements for Fleet Numerical to obtain this model for testing in the North Pacific. She adapts it to be initialized from the NOGAPS model. She renames the model GFDN (N for Navy), so it may be differentiated from GFDL.

However, GFDN does not do well with the NOGAPS initialization. Mary Alice traces the problem to a too dry stratospheric layer in NOGAPS. After meetings with NEPRF global model personnel, who initially designed NOGAPS and generate improvements of this model for Fleet Numerical, the modeling in the stratosphere is appropriately modified to allow higher moisture content.

GFDN eventually becomes the best tropical cyclone forecasting model for the North Pacific. In addition, the accuracy of NOGAPS, especially in the stratosphere, is also improved. Had it not been for the support Fleet Numerical is providing in the tropical area for JTWC, this dryness would have persisted for a number of years.

My software extracts tropical cyclone positions from NOGAPS forecast fields to support JTWC, but at this time the global model usually does not do as well as the GFDN regional model with such a small-scale feature as a tropical cyclone, even after it becomes a typhoon.

My biggest Cold War accomplishment is my work to support the Submarine Launched Ballistic Missile (SLBM) effort. I can't tell you what I am doing, but I am between a rock and a hard place on solving a particular problem. At this time Don Chin (GSA onsite contractor), who had been working for me at SASC/STX, is having a problem and asks me to come over to his office for assistance in determining a solution. I do, and we solve his problem.

Upon returning to my office, I realize that with a little bit of tweaking here and there I can use this solution to solve my problem! I write of this, because this is the way a lot of progress is made in scientific software development. This is one reason there are American Geophysical Union presentations – for the effective cross-pollination that occurs!

Oddly, about two or three Christmas Eves in a row I have to go into Fleet Numerical at night to rectify a problem with my SLBM software! These are the only times this program ever had any problems.

As requested, I also write oceanographic programs for Leo. I no more than finish a new global ocean surface currents model and a revised search and rescue program, than a navy pilot bails out of his jet off the west coast and the San Francisco Coast Guard Unit requests Fleet Numerical support, as they usually do. The search and rescue program uses the currents from the global surface currents model and the forecast surface winds from the atmospheric global model (NOGAPS) to predict his location. The pilot is located right where my software predicts. This makes me feel good. The pilot would probably vote this my most important work!

19
SUMMATION

So, I became a Midshipman in the NROTC program to help pay for my college, became a Naval Aviator because I wanted to learn to fly and by happenstance, perhaps starting with my requesting a west coast squadron or by my assignment to the *only* Skyraider squadron on the west coast that is not an attack squadron, became a Meteorologist. However, it has turned out to be very good for my family and me. All of my assignments since VAW-11 prepare me to be a successful navy meteorologist and a profitable civilian manager/meteorologist!

I am challenged by both my navy and civilian work. I also enjoy working for the Navy, Systems and Applied Sciences Corporation, STX and various other companies under Government Services Agency contracts. We have no more changes in location, Monterey, after the summer of 1969 and no more household moves after June 1972, which are a definite plus.

However, without the great support I received from Jackie over the years I could not have accomplished what I did! She deserves a lot of credit for my successes!

I start my civilian retirement on Wednesday, 1 October 1997.

Post Scripts

Several of my former employees, who are retired naval officers, Don Chin, Bud Hinson, Ralph Sallee, Carol Strey, and I have lunch most Fridays at the Naval Postgraduate School.

Over the years, the main effort to improve tropical cyclone forecasting has been to develop the *best* regional tropical cyclone forecast model! A lot of money and effort have been expended to accomplish this!

At the present time, the *best* "model" to accomplish this is the average of all the numerical prediction models, global and regional. "All" means all the viable numerical models in the world, which are about 8 or 9 for the western Pacific. Using a simple average of these models is producing the most accurate forecasts, from a statistical annual approach. Removing the worst forecast from the previous forecast or weighting the best forecast from the previous forecast makes it worse than the simple average!

I find the above very interesting; hope you do also.

* * * *

Which airplane followed the AD-5W for Airborne Early Warning (AEW) services in the Navy? As a stop-gap effort the Grumman WF-2, known as the "Willy Fudds," were initially developed; but never became operational. The E-1B Tracers became operational in the late 1950s in VAW-12 and early 1960s in VAW-11. The radar was the improved APS-82, developed by Hazeltine. This equipment was operated by two radar specialists. When on station, the copilot left his normal position to sit on a fold-out seat between these two specialists. He assumed the duties of tactical coordinator. This plane had a wing span of 69 feet 8 inches and was powered by two Wright R-1820-82WA Cyclone radial piston engines. Each engine had 1,525 hp. Starting in 1964 the E-1s were replaced by the Grumman E-2 Hawkeye in VAW-11 and the following year in VAW-12. The radar has been upgraded from APS-138, to -139 to -145 by 1991. The power plants are Allison turbo-prop engines – earlier version had 4,600 shp per engine; but this has been upgraded to 4,910 shp each. The wingspan is 80 feet 7 inches, length 57 feet 7 inches. Improved versions of this plane are still being used today.

Douglas AD-5W, Grumman E-1B and E-2 Hawkeye (US Navy Official)

(Note, in front of each aircraft are crewmen – one pilot and two radar specialists, two pilots and two radar specialists and two pilots and three radar specialists.)

Likewise the aircraft carriers have become much larger. On 7 September 1968 the USS *John F. Kennedy* (CV-67) entered service, as shown on next page, and since then the carriers are still becoming larger.

USS *John F. Kennedy* (CV-67) (US Navy Official)

(Note the two E-2As (wings folded) next to the island.)

She has four elevators, catapults and arresting wires. Beam of 128 feet, but flight deck is 267 feet wide. Overall length is 1050 feet. She draws about 36 ½ feet. Her displacement is approximately 80,950 tons. Speed 30+ knots.

Last Flight Story

On 4 October 2003 at the Marina, California annual Air Faire I finally stop procrastinating and ride in an open cockpit biplane, shown below. This aircraft was a training plane, a N3N-3, built by the Naval Aircraft Factory. Their nickname is *Yellow Peril*. These fabric-covered planes were built in late 1939 till 1942, are powered by a 235 hp Wright R-760-2 radial swinging a two-bladed, fixed-pitched wooden propeller and have a top speed of 109 kts. They measure 34 feet in span, 25½ feet in length, weigh 2,100 pounds empty.

Naval Aircraft Factory N3N-3 (By author)

I fly in the front seat wearing a cloth helmet and goggles! All that is missing is my white scarf; which, unfortunately, is packed away at home. It is a great thrill for me to experience this old-time way of flying! The pilot, Garry Hendrickson, who is also a former naval aviator, lets me fly his plane (shown above), which is light on the controls and maneuverable. It is fun hearing the wind whistling by the wings, struts and fuselage and feeling the air rushing past my head and shoulders!

When I return home, Jackie is not surprised that I had gone flying in a bi-plane this day. I guess she knows pretty well after nearly fifty years of marriage.

With this I've finally flown in all basic conventional plane types – single-engine, twin- engine and four-engine; propeller-driven and jet-driven; land-plane, sea-plane and ski-plane; powered and glider, single-winged and bi-winged; high-winged and low-winged; and open-cockpit and closed-cockpit. And I've flown in a helicopter, for good measure.

My first flight was in a light, high-winged plane over Medford, Oregon when I was 5 years old. I was fascinated with the view – a fascination I never lost.

Author sitting in his favorite WWII aircraft, Spitfire Mk IX, several years ago, also in Medford, Oregon.

20
ADDENDUM

The following is from: Air Test and Evaluation Squadron 23
Fall 2001 Projects:

Joint Precision Approach Landing System (JPALS)

Imagine after the break at the ship, you drop the gear, flaps, and hook, and then while still on downwind, engage the autopilot for an automatic hands-free Case I coupled approach. It is just one of the many capabilities of the future joint precision approach landing system (JPALS). In April 2001, the JPALS test team successfully performed the first global positioning system (GPS)-based automatic landing to an aircraft carrier aboard the USS *THEODORE ROOSEVELT* (CVN-71).

Salty Dog 110 rolling into position to trap aboard USS *Roosevelt* (CVN-71)

Based on GPS, JPALS is intended for military aircraft including manned and unmanned fixed-wing, vertical takeoff and landing (VTOL), and rotary-wing aircraft. JPALS is designed to replace tactical air navigation (TACAN) systems and augment the current automatic carrier landing system (ACLS) and instrument carrier landing system (ICLS). Other potential capabilities of JPALS include:

- Autopilot flown departures and s
- Automatic recoveries from 200 miles from the ship (no more holding pattern?)
- Automatic landings at divert or expeditionary airfields
- Two-way data link between aircraft and ship/shore stations for voiceless communication between pilot and controller of information such as traffic, current fuel states, bingo fuel states, recovery instructions, and divert status

JPALS is a flexible system for multiple platforms and will help redefine the current military concept of operations in the next decade.

Improved Fresnel Lens Optical Landing System (IFLOLS)

The initial production of IFLOLS is in full swing with three production systems installed so far aboard the USS *Stennis*, USS Nimitz, and USS *George Washington*. For those of you who have already flown IFLOLS, you know what a great system improvement it is with 12 fiber optic cells (the central vertical row of lenses) indicating glide slope at about twice the sensitivity as the older FLOLS – which became available in the early sixties. With IFLOLS's greater definition, it is possible to accurately reference the lens as far out as a mile behind the ship. The greater sensitivity of the lens cells provides the pilot the capability to observe glide slope deviations sooner, thus promoting smaller corrections and tighter control of glide slope. Once you recalibrate your eye to flying on the new system, it's tough to go back to regular FLOLS. For those of you who have not seen IFLOLS yet, you are in for a treat.

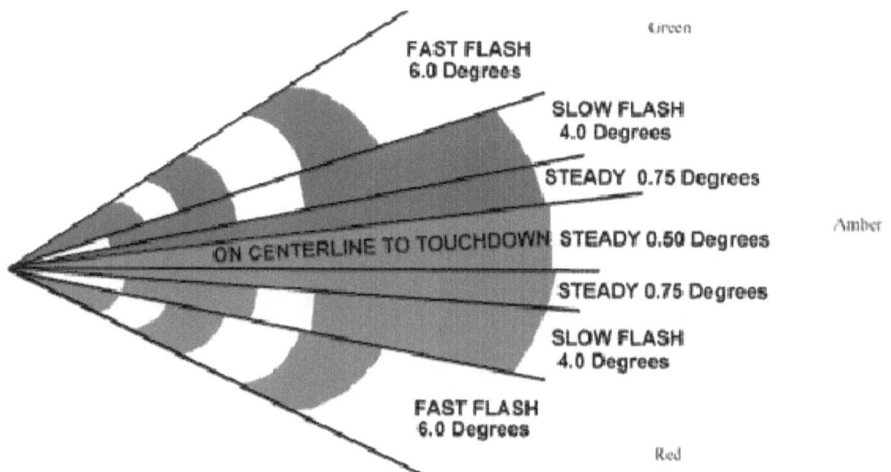

Long-Range Lineup System (LRLS)

Squaring away lineup early is critical to flying a good pass. Beyond a mile or two from the ship, unfortunately, the carrier box and drop lights appear too small to be compelling enough for maintaining lineup precisely. LRLS was designed to help aviators capture and maintain centerline out a range during straight-in approaches until inside a mile from touchdown, where the drop lights and carrier box become effective visual cues. The LRLS (shown above in plan view) uses green (starboard), amber (center), and red (port) lasers mounted on the fantail and pointing aft along the extended centerline of the angled deck. When the aircraft is on centerline, LRLS appears amber. When the aircraft deviates left or right of centerline, LRLS appears red or green, respectively. If the aircraft deviates even further from the centerline, LRLS appears to flash.

The further the aircraft is from centerline, the faster LRLS appears to flash.

Installation of LRLS is initially planned to closely follow the installation of IFLOLS.

APPENDIX

Technical Information

For those of you interested in the engineering/operational aspects of the flight stories, the following discusses the reciprocating engine controls a pilot uses and their interaction while flying a Douglas Skyraider.

The propeller governor sits on the end of the propeller shaft and monitors the RPM of the engine with a centrifugal device. If the RPM starts to increase the governor very slightly increases the pitch of the all the blades simultaneously. Likewise, if the RPM starts to decrease the governor similarly decreases the pitch. Thus, for all practical purposes the RPM is kept constant. An "over-speed" occurs when the governor fails and all the blades automatically go into flat (near-zero) pitch – the sudden lack of a normal amount of air resistance by the propeller causes the engine to instantly increase in RPM – above the redline limit, if not caught *immediately*.

The throttle controls the air intake manifold pressure, which is measured in inches of mercury by a dial on the instrument panel. There is a limit as to how high the manifold pressure may be for any given RPM setting. The "prop" is the name of the RPM-setting lever. If you increase the manifold pressure above this limit, you "over-boost" the engine – and it will start coming apart – literally! On takeoff, you place the prop control in its maximum RPM setting – all the way forward, which is the lowest pitch setting. Then you push the throttle all the way forward during the takeoff run (or while you're retained by the catapult hold-back ring).

The throttle is set such that you cannot over-boost the engine by placing the throttle all the way forward. Whenever you reduce the

223

power of the engine, you *always* pull the throttle back first; then you adjust the prop setting for climb, cruise, or whatever.

Before you increase the throttle above the "over-boost" limit, you *always* increase the RPM setting first for the manifold pressure you want to have. Then you increase the manifold pressure by advancing the throttle.

If you get these sequences reversed, you don't fly long!

You should memorize these manifold pressure limits for each RPM setting – or at least have them written on your kneeboard, for they are not posted in the cockpit!

I have been told that during primary training, the instructor *always* held his hand behind the prop control to make sure the student did not pull the prop control back before the throttle after a practice takeoff! And likewise the instructor *always* prevented the student from accidentally increasing the throttle before the prop control when initiating a climb.

In addition, the AD throttle quadrant contains the mixture-control - you takeoff and land in "Rich" and fly in "Normal" mixture settings. A high-powered climb also requires a "Rich" setting. The "Rich" setting allows extra cooling for the engine, as well as more power. The third position is "OFF." However, on *very* rare occasions some old time pilots would sneak it below "Auto-Lean," (same as our Normal) but this is *very* risky. It is too easy to burn the engine valves with a too-lean mixture.

During WW II most planes did not have "Auto-Lean" and pilots did adjust the richness, as required, to improve their flight mileage.

The final item on the AD throttle quadrant is the friction control knob – which you *always* have set very tight for a catapult shot and an arrested landing!

* * * *

Once I was standing near "Pri-Fly" on the island of an aircraft carrier overlooking the flight deck while some ADs were conducting carrier landing practice – arrested landing followed by a deck launch before the next plane came aboard. I watched an AD complete its arrested landing, the deck personnel got the wire detached from the tailhook, and then the plane taxies forward a short distance while aligning itself with the forward straight section of the flight deck before stopping; by following taxi signals.

The high-power run-up signal was given to the pilot, followed quickly by the launch signal.

As the pilot pushed the throttle from high-power to full power during the takeoff run, I hear the Air Boss *scream* over his bullhorn, "*RICH, RICH, RICH!*" He and his associates could tell by the *sound* of the engine, as it was going from high-power to full power, that the pilot had *not* set his mixture control to "Rich." This meant he had failed to set it to "Rich" prior to arriving at the 180-position for his approach and had not double-checked his engine controls prior to the high-power run-up preceding the launch.

GLOSSARY

AC. Aircraft Controller, in front right seat of AD-5W

AD. Douglas Skyraider operational aircraft - later designation **A-1** basic version is an attack aircraft

AEW. Airborne Early Warning

APT. Automatic Picture Transmission of weather satellite picture

ASW. Antisubmarine warfare

ATC. Air Traffic Control

ATG4. Air Task Group 4

BB. Battleship

CAPT. Naval officer, rank of captain

CarQuals. Aircraft carrier qualifications (pronounced: care-quals)

Cat Shot. Launch aircraft via catapult, hydraulic or steam powered

Cold Cat Shot. Force of catapult fails to get aircraft safely airborne

CDR. Naval officer, rank of commander

CIC. Combat Information Center

CO. Commanding Officer

COMNAVAIRPAC. Commander, Naval Air, Pacific

COMNAVMAR. Commander, Naval Forces, Marianas (Islands)

CV. **(C)** Aircraft carrier of **(V)** heavier-than-air aircraft

CVA. Attack aircraft carrier

CVL. Aircraft carrier built upon a cruiser hull

CVN. Nuclear powered aircraft carrier

CVS. Antisubmarine aircraft carrier

DD. Destroyer

DM. Destroyer Minelayer

DR. Dead Reckoning navigation method

ENS. Naval officer, rank of ensign

FAA. Federal Aviation Administration

FAMOS. Fleet Applications Meteorological Observations Satellites

FCLP. Field Carrier Landing Practice

FWC. Fleet Weather Central

G's. Force of gravity: 3G's – 3 times the force of gravity

GQ. General Quarters – battle stations

Hp. Horse power

IFR. Instrument Flight Rules

JTWC. Joint Typhoon Warning Center

Kts. Nautical miles per hour; 1.15155 miles per hour

LCDR. Naval officer, rank of lieutenant commander

LT. Naval officer, rank of lieutenant

Lt. Col. Air force officer, rank of lieutenant colonel

LTJG. Naval officer, rank of lieutenant junior grade

LSO. Landing Signal Officer

MM. Martin Marietta

NOGAPS. Navy Operational Global Application Prediction System

NROTC. Naval Reserve Officers Training Corps

O-in-C. Officer in Charge

OOD. Officer of the Deck

OPARS. Optimum Path Aircraft Routing System

ORI. Operational Readiness Inspection

Pri-Fly. Primary Flight control compartment on the port side of the island of an aircraft carrier. Where the Air Boss controls all flying aircraft near the carrier and the movement of aircraft on the flight deck.

MLS. Mirror Landing System - Gyro-stabilized lens system.

NPS. Naval Postgraduate School

RPM. Revolutions per minute

SAC. Strategic Air Command, Air Force

SDO. Squadron Duty Officer

Shp. Shaft horsepower

SNAFU. Situation normal, all fouled up

SNB. Beachcraft twin-engine navy utility aircraft

SNJ. North American navy flight training aircraft

SOP. Standard operating procedure

SS. Steam ship

T-28B. North American navy flight/instrument training aircraft

Typhoon. Tropical cyclone in the western pacific with sustained wind speeds of at least 65 kts (74 miles per hour).

USS. United States Ship

UT. University of Texas, Austin, Texas

UHF. Ultra high frequency

VFR. Visual Flight Rules

www.ingramcontent.com/pod-product-compliance
Lightning Source LLC
Chambersburg PA
CBHW031836170526
45157CB00001B/318